IMAGES
of America

KNOXVILLE'S WIVK

ON THE COVER: This is a promotional photograph taken of a pony giveaway outside the WIVK studios on Gay Street. (Photograph courtesy McClung Historical Collection, Knox County Public Library System, Knoxville, Tennessee.)

IMAGES
of America

KNOXVILLE'S WIVK

Ed Hooper

ARCADIA
PUBLISHING

ISBN 978-0-7385-6703-7

Published by Arcadia Publishing
Charleston SC, Chicago IL, Portsmouth NH, San Francisco CA

Printed in the United States of America

Library of Congress Catalog Card Number: 2008930683

For all general information contact Arcadia Publishing at:
Telephone 843-853-2070
Fax 843-853-0044
E-mail sales@arcadiapublishing.com
For customer service and orders:
Toll-Free 1-888-313-2665

Visit us on the Internet at www.arcadiapublishing.com

This book is dedicated to Jim Dick and the entire WIVK family.

Contents

ACKNOWLEDGMENTS

This book was truly a community effort. The photographs in this book were graciously supplied by WIVK, Ed Brantley, Dave Young, Steve Cotham at the East Tennessee Historical Society's McClung Collection, Foster Arnett, and Jean Ash. Unless otherwise noted, photographs in this book were supplied by unofficial station archivist David Young. Supporting material and research contributors include the above and F. A. Blankenship, Jan Oliver, Bobby Denton, Sheila Hudson, Alan Williams, University of Tennessee Hospital senior vice president Norman Majors, Connie Payne, and Ted Ousley.

INTRODUCTION

Knoxville's WIVK is an American landmark in the broadcast industry. The country music radio station has never strayed far from its roots and has often rewritten the book on the business. It is a fitting tribute to founder James A. Dick, who created the station call letters to stand for "We're the Independent Voice of Knoxville."

While the city is home to some of the earliest radio stations in the nation, Knoxville is one of the most unique broadcast markets in America. The broadcast industry is a complex world where there are many pretenders to the throne, and normally that is how it works no matter where you live. Ratings juggle stations from one position to another, and these businesses prosper and sometimes fail based on those numbers.

Knoxville's WIVK is different. Its origins, development, evolution, and that independent voice has made it one of the greatest broadcast success stories in the United States, but it is more than music that made it the No. 1 country music radio station in the nation. As this book is being written, WIVK goes completely against the norm for the industry. It has only had two owners, three general managers, and three morning shows since it signed on the air in the 1950s. Its demand for excellence among its employees and their dedication and loyalty to both their respective jobs and the station is legendary in broadcast circles. It has launched many great careers in news and entertainment, but it is probably best known for the sheer number of country music celebrities that got their start on its airwaves.

With leadership and prosperity come civic responsibility, and this is a point driven home by WIVK's service to Knoxville and East Tennessee. The station has set a remarkable standard that has earned it national recognition with its dedication to news, information, and participation in numerous causes that aided the region and their listeners. The news department holds as many prestigious awards as does its drive-time announcers and has sustained radio news in an era where television and the Internet have virtually wiped it out in local markets. As a reporter covering news in Knoxville for many years, I know from experience you could always count on seeing the WIVK microphone flag in a newspaper photograph or a television shot at a press conference. Their coverage of breaking news has often been picked up and carried by national news networks. Sports broadcasting was another landmark for WIVK, especially when it came to the University of Tennessee (UT) and football season. The station analysis of the games and their weekly broadcasts are some of the most listened to in the nation when the "Big Orange" hit the field in the fall of the year. Through thick and thin, the station has always had the best coverage and helped propel UT sports to the place it holds now in East Tennessee homes. Television did not change how the game was delivered that much. Many still watch it on television with the sound off and WIVK radio doing the play-by-play.

The same can be said for their commitment to public affairs. Even when Federal Communication Commission regulations were relaxed on the topic, WIVK maintained its vigilance and sustained itself as a leader in public affairs broadcasting, believing in the concept of full-service radio when most stations abandoned it to go to satellite formats and an advertising sales staff. But it always comes back to the music, the listeners who turn the dial, and the announcers presenting day-to-day programming that have made WIVK the most listened to country music radio station in America—one still sought out by country music entertainers, celebrities, and local and national leaders in both broadcasting and government. It has been said that familiarity breeds contempt, but, in this case, it bred a No. 1 radio station that is still going strong.

This book is a look at the colorful story of that station and founder Jim Dick's idea that blossomed into an industry standard few broadcast organizations can emulate. As an alum of the station myself, I can state there are no smoke and mirrors here. It is a radio station run by the old school on a hard-earned formula that not only works, but also dominates an entire region of Tennessee in entertainment, programming, and advertising. This is the history of Knoxville's WIVK.

One

WIVK Signs on the Air

James Allen "Jim" Dick was born on August 27, 1919, and spent his early years growing up in Paris, Tennessee. He gained an interest in radio early in his life because of the work of his uncle Felix Holt. His uncle was a writer who had penned scripts for the Lone Ranger radio series and had written a number of novels—one was *The Gabriel Horn*, which was made into a feature film called *The Kentuckian* starring Burt Lancaster and debuting a new actor named Walter Matthau.

Jim Dick developed a love for music early in his childhood and learned to play piano, which he did in bands in his teenage years. He was an accomplished piano player and developed a love for the instrument, even building one on his own when he needed an instrument to play.

In 1941, he was drafted into World War II, where he was assigned to the horse cavalry and later transferred to Cadillac's tank maintenance school. From there, he was sent to the U.S. Army base in Carmel, California, and trained on the landing ship tanks being designed. When the war ended, Jim quickly headed back to his Paris, Tennessee, home. Like most returning soldiers of the day, he took advantage of the G.I. Bill and attended Transylvania University in Kentucky.

While going to school, he got the chance to start working at a local radio station sweeping floors and doing odd jobs. He was working there one day when Clarence Beaman of Knoxville's WKGN called the station to offer a sales job to one of Dick's colleagues at the station. The colleague turned it down, but Jim Dick said he was interested in the job, and the manager put in a good word for Jim as a sales representative.

During this time, he met Marilyn Mandle, who was the younger sister of a friend of his, and soon the two fell in love and were married on March 18, 1947. In August of that year, he and Marilyn decided to take their chances and move to Knoxville, where Jim went to work at WKGN. He explained to the manager that he did not really have that much sales experience, but soon Jim Dick became a pretty good advertising salesman for the station.

After a short time at the station, he left to go to work for WBIR and excelled at his job selling advertising. He developed a close friendship with WBIR founder John Hart. Jim Dick worked himself through the ranks of the station to become commercial manager at WBIR radio. He then decided to take a chance at a lifelong dream.

In 1952, at the age of 32, Jim Dick applied to the Federal Communications Commission (FCC) for a permit to take over a daytime-only, 1,000-watt broadcast signal at AM 860 that had been vacated by another station. In the 1950s, when the airwaves were not so cluttered, it was a powerful signal that covered the region. The only immediate problem was that the FCC moves slowly, and it was a year before Jim Dick saw the license necessary to sign the station on the air.

Those days were also an era of propriety and strict licensing requirements in broadcasting. It was regarded as unethical for Jim Dick to work for one radio station while applying for a permit to own another one. From the time he filed for the vacated signal until he received the license, Jim Dick held a variety of jobs in Knoxville selling clothing, shoes, or whatever to make ends meet. He managed to do pretty well and put away enough money to begin working on a broadcast studio at 319 North Gay Street in downtown Knoxville. The new station was on the second floor of Greens Hardware Store across from the Regas Restaurant. Jim Dick was there for every nail driven and piece of lumber sawed.

It was not just the building and the studio Jim Dick was constructing. He began to look for the talent and people needed to operate the station. His experience with the business of broadcasting included having an eye for talent, and so he began stocking the station and starting the staffing process.

Along with program director Jim Hess, Jim Dick hired a young man still gong to UT to help sort records and do odd jobs. Like Dick, the young man had gotten out of the army and come home to study, but he too had a love for radio and was willing to work to help launch the station properly. On March 20, 1953, all the construction was over, and the young man, named Claude Tomlinson, put on his best game face and opened the microphone. "This is WIVK—Knoxville's newest radio station signing on the air for the first time." Those were the first words spoken by Tomlinson. With six employees, a few pieces of equipment, and little money, WIVK and Dick Broadcasting were born.

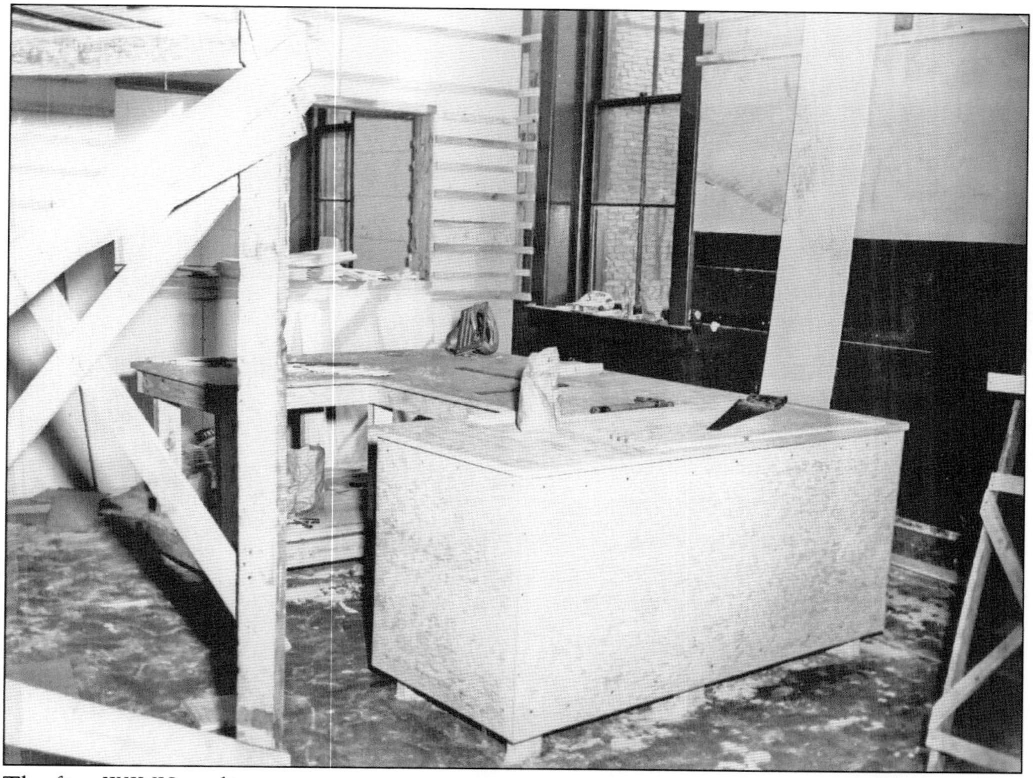

The first WIVK studio was a rented second floor located at 319 North Gay Street. This is the control room construction that kept Jim Dick busy while he worked in a number of jobs to pay the bills and build the studio.

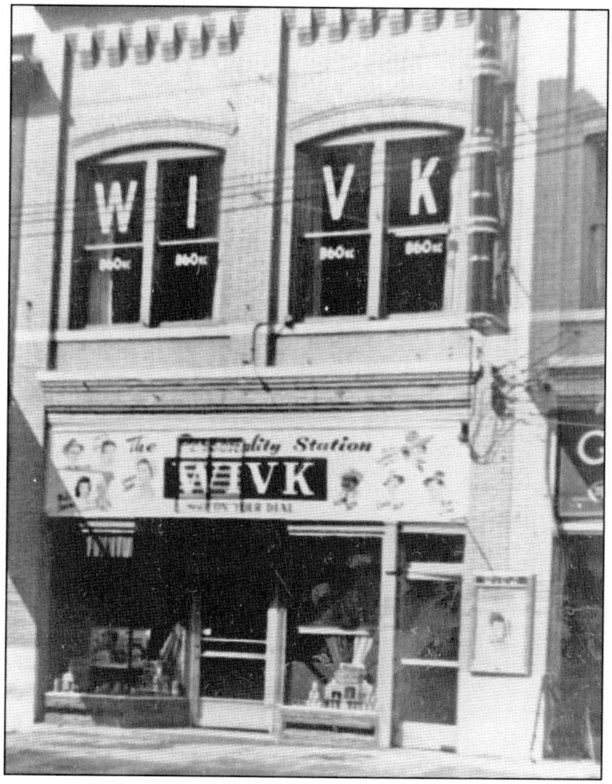

This is the storefront on Gay Street. Although a hardware store was on the bottom floor, the station logo dominated the building. It was from here that the station would take on the powerful WNOX, which was a 10,000-watt AM station that ruled the city. (Photograph courtesy WIVK.)

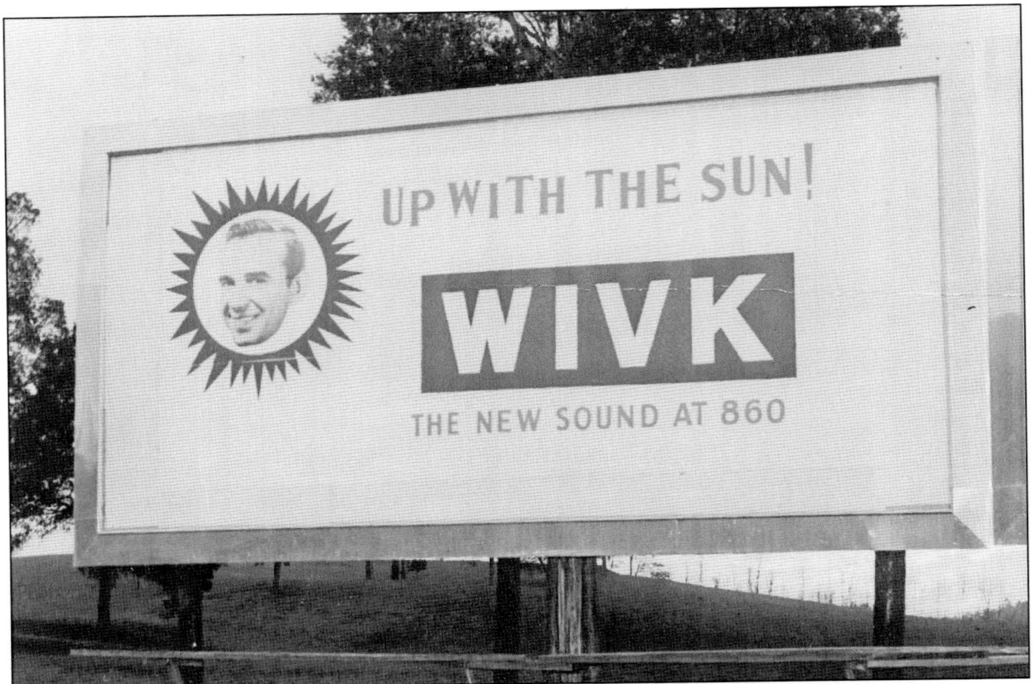

This station billboard shows the first WIVK frequency at 860 AM on the dial. It would later be changed to 850.

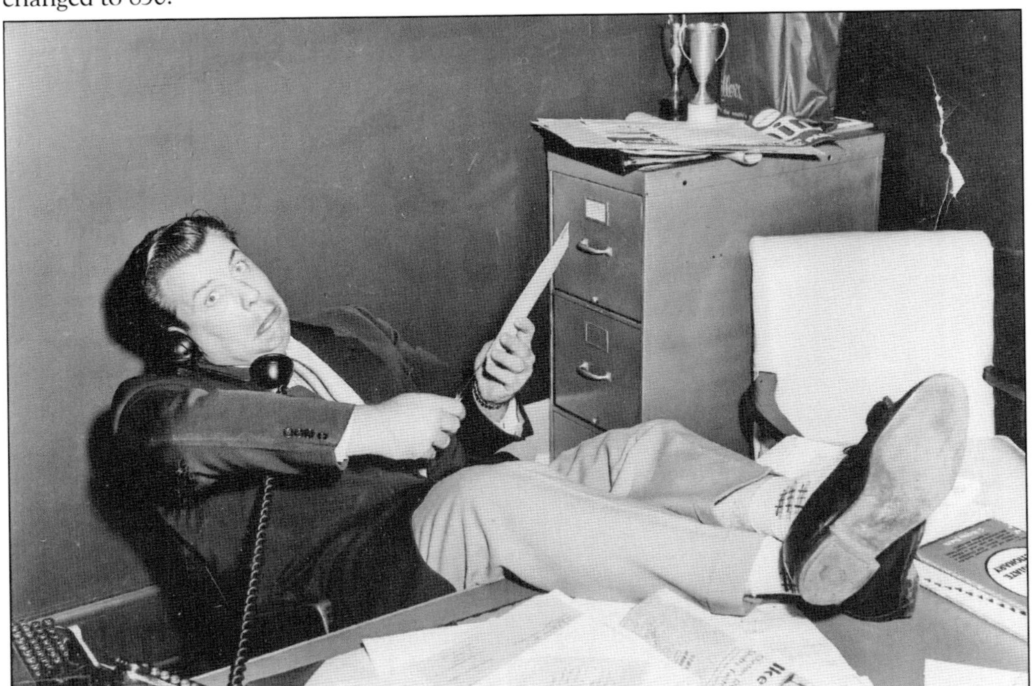

The first program director of WIVK was Jim Hess, who everyone would come to know as "Big Jim." Jim Dick and Jim Hess began the station in the age of radio broadcasting that used "block programming," which varied in format, and each show would have its own following of listeners.

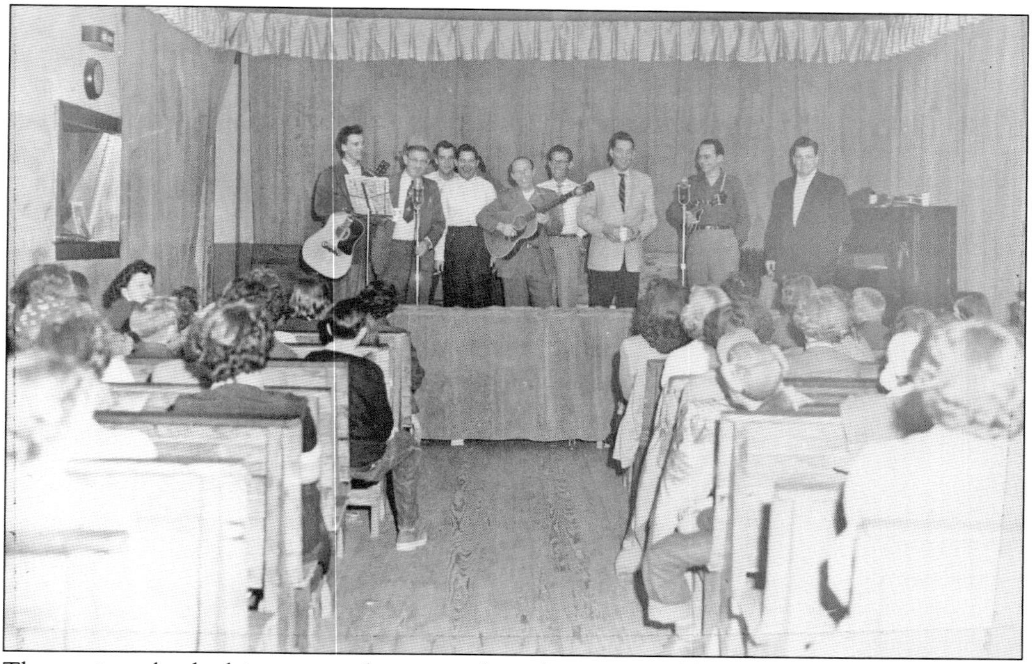

The station also had its own auditorium, where listeners could hear live bands perform and watch the radio programs as they were broadcast. This photograph features colorful local grocer Cas Walker (standing, second from left). Standing third from left is 1950s country music star Carl Butler. (Photograph courtesy WIVK.)

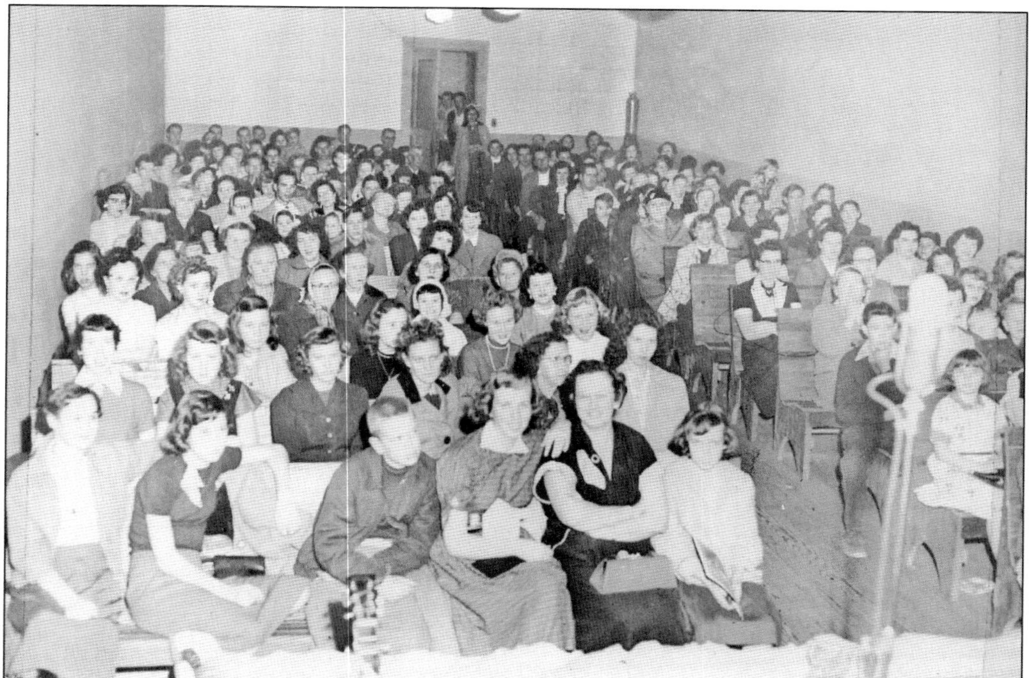

The auditorium, pictured here looking from the stage, shows a respectable crowd for the WIVK program. This was a common sight, especially when popular entertainers would visit the station to perform, as was the rule of the day. (Photograph courtesy WIVK.)

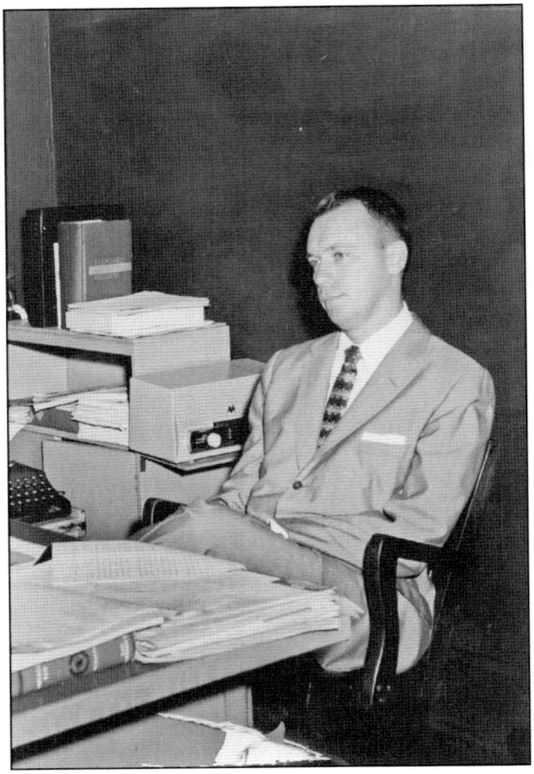

This is a photograph of WIVK founder Jim Dick taken in the early 1950s. The station's early success was due to his hands-on efforts in sales and making sure the Independent Voice of Knoxville stayed true to its name.

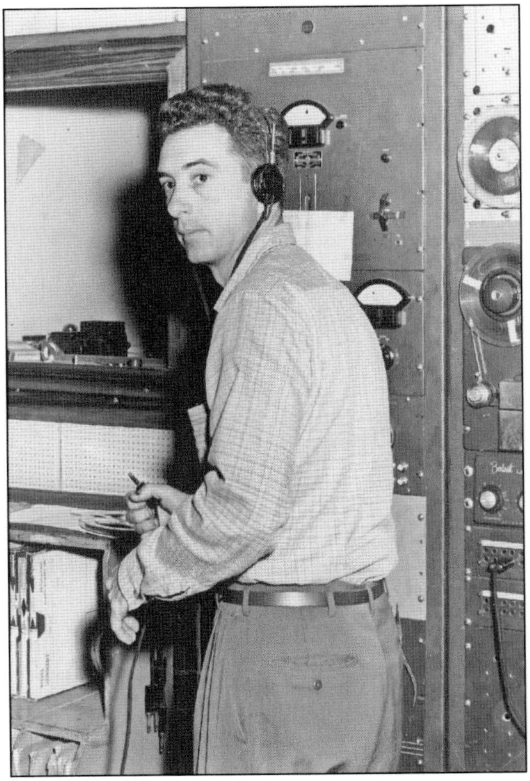

Any broadcast station is only as good as its engineer. This is a photograph of WIVK's Milton Jones. He signed the station on the air for the first time and remained with it for the rest of his career.

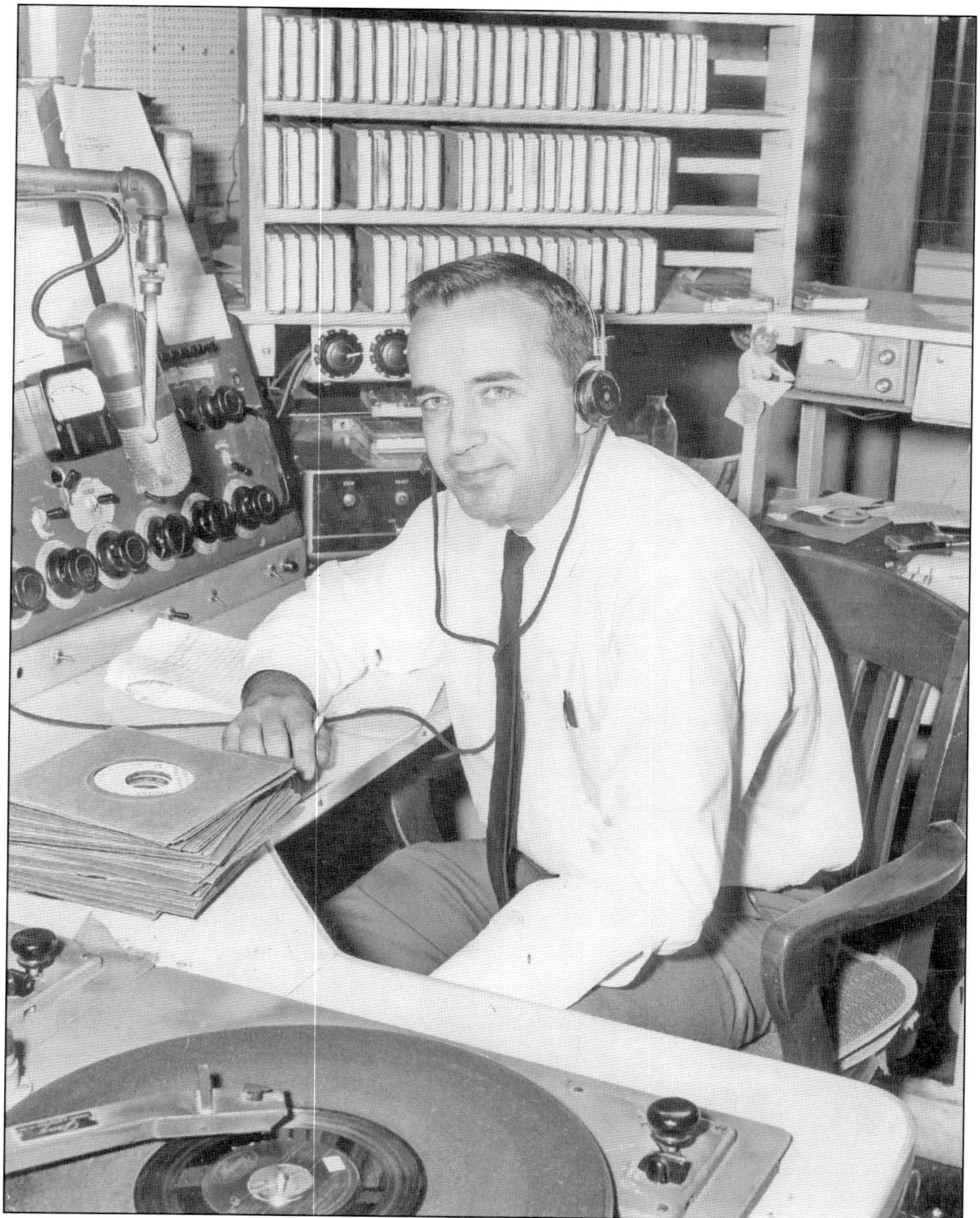

When not playing his characters, Claude Tomlinson was a first-rate broadcaster. He was attending the University of Tennessee and working at the college station when he saw the job opening posted for WIVK. He was hired to sort record albums at first and signed the station on the air for the first time on March 20, 1953. His first words were "Good morning. This is WIVK—Knoxville's newest radio station signing on the air for the first time." (Photograph courtesy McClung Historical Collection, Knox County Public Library System, Knoxville, Tennessee.)

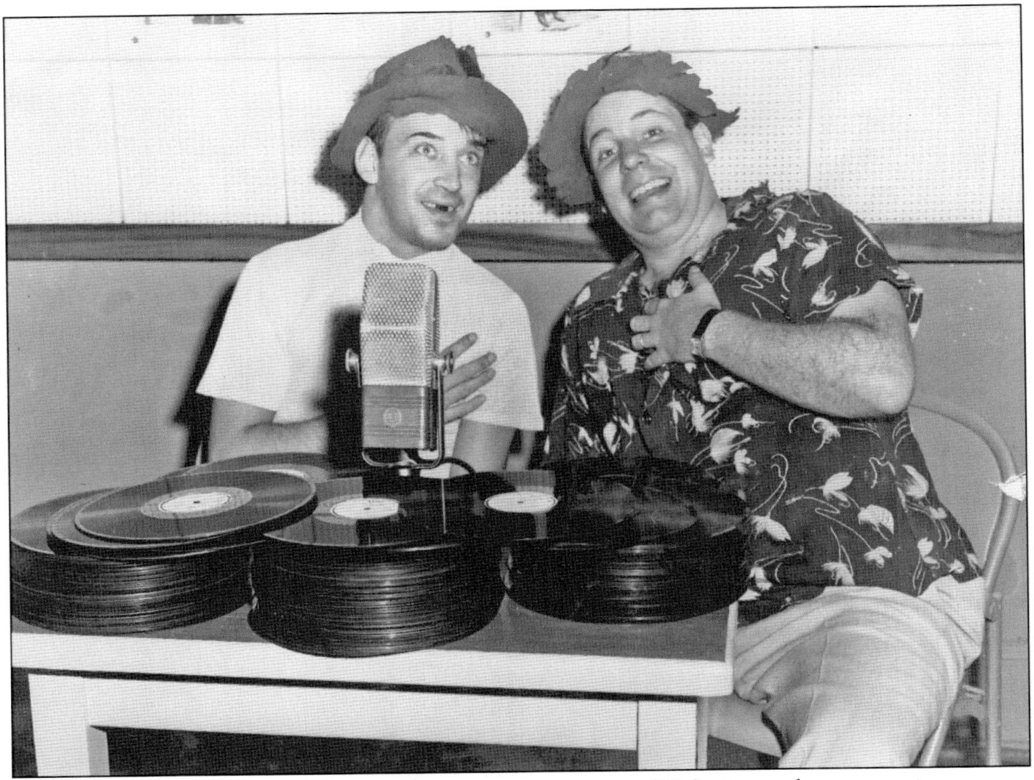

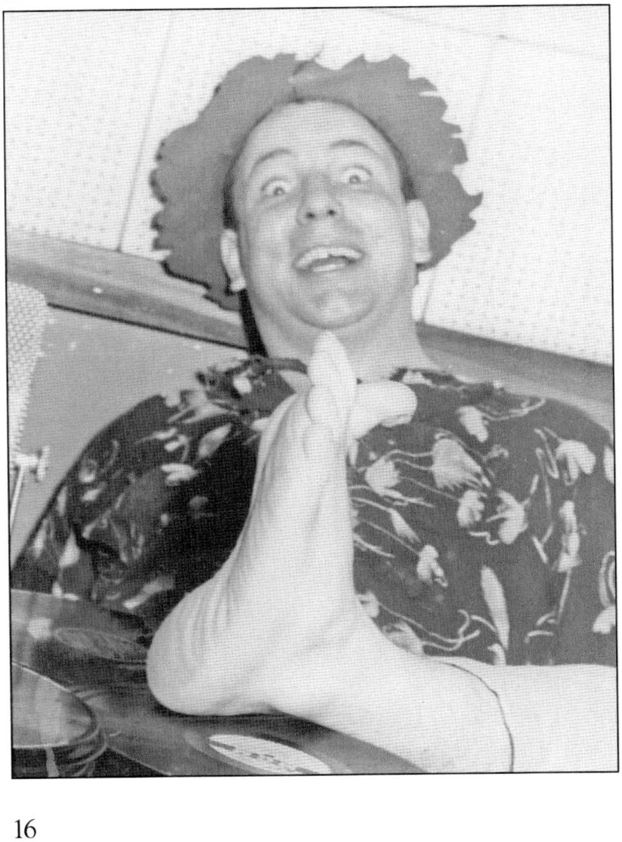

With six employees starting out, program director Jim Hess (right) doubled in sales and on-air work with the legendary Claude Tomlinson (left). This is a photograph the characters "Big Jim" and "Little Alf," who did the morning program.

"Big Jim" Hess was a radio personality for years on WIVK before moving on to local television. He made a name for himself as a commercial pitchman and as a wrestling announcer in Knoxville.

JFG COFFEE COMPANY, INC.

P.O. BOX 2151 KNOXVILLE, TENNESSEE
Phone 5-3176

IMPORTERS HIGH GRADE COFFEES ROASTERS

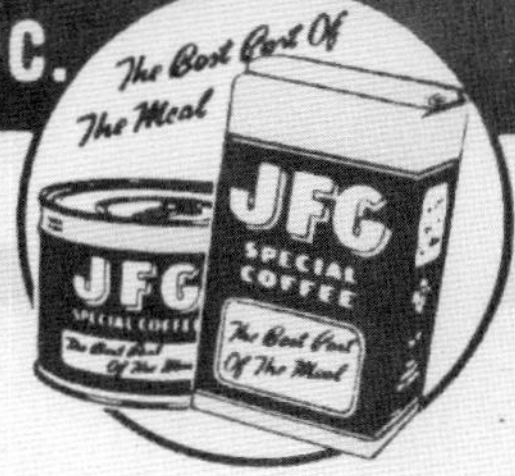

August 18, 1956

Radio Station WIVK
Gay Street
Knoxville, Tennessee

Attention: Mr. Jim Dick

Dear Jim:

The results of our recent radio check on four Knoxville stations has been completed and I thought you would be interested in the results.

Station WIVK lead in the total number of responses and also in the different number of towns and localities. Some responses were received as far away as Kentucky, North Carolina, and Georgia. These results I guess bear out pretty well the last major survey made of the Knoxville market.

With respect to cost, we also used the television station here and while the response was three times as great as the leading radio station, the cost per response was five times greater.

Alterations in our radio schedule have been made so as to coincide with the results of this recent test.

Thanking you for your cooperation and the fine job you are doing for us, I remain

Cordially yours,

R. E. Snow, Jr.

R. E. Snow, Jr.
Advertising Manager

RES:by

This is a letter to Jim Dick from the JFG Coffee Company, Inc., regarding a survey done in 1956, which shows that the station was the leading broadcast advertiser in East Tennessee.

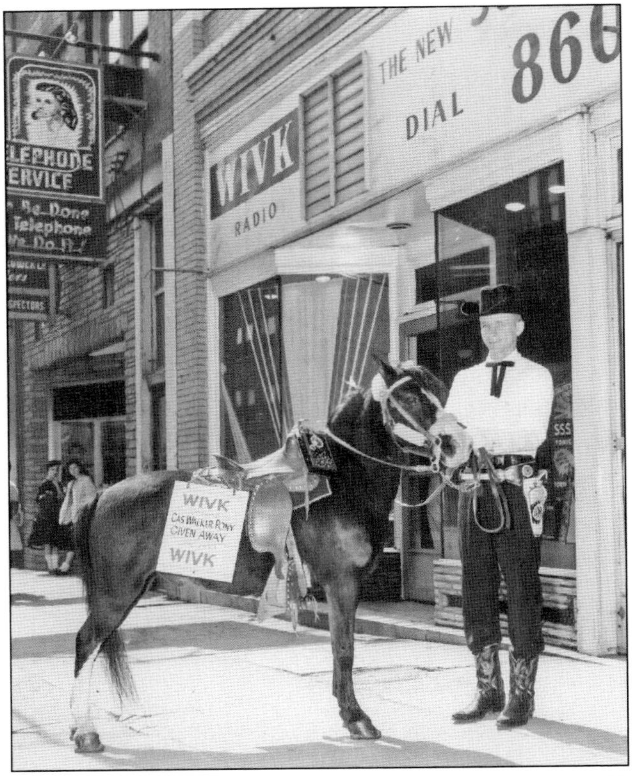

East Tennessee grocer Cas Walker was a prolific promoter of his grocery stores. Here is a promotional photograph taken of his "Pony Give Away." (Photograph courtesy McClung Historical Collection, Knox County Public Library System, Knoxville, Tennessee.)

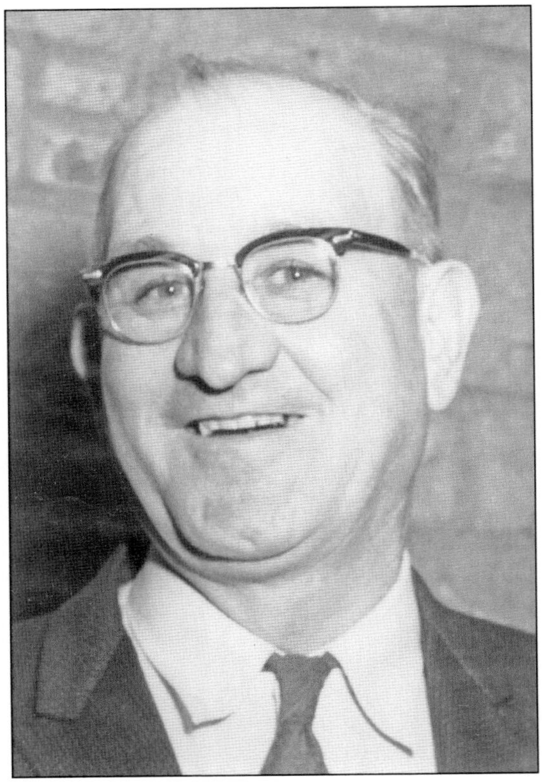

Cas Walker was more than a grocer in Knoxville. Many called him the musical grocer because of his close ties to local and regional musicians who later rose to become stars. Among them were the many Grand Ole Opry acts and country music superstars like Loretta Lynn and Dolly Parton. He was also an active political figure whose antics were often written about in the press on local and national levels. He maintained broadcast programs on WIVK and later on WBIR-TV that always stayed true to his love of country music. (Photograph courtesy McClung Historical Collection, Knox County Public Library System, Knoxville, Tennessee.)

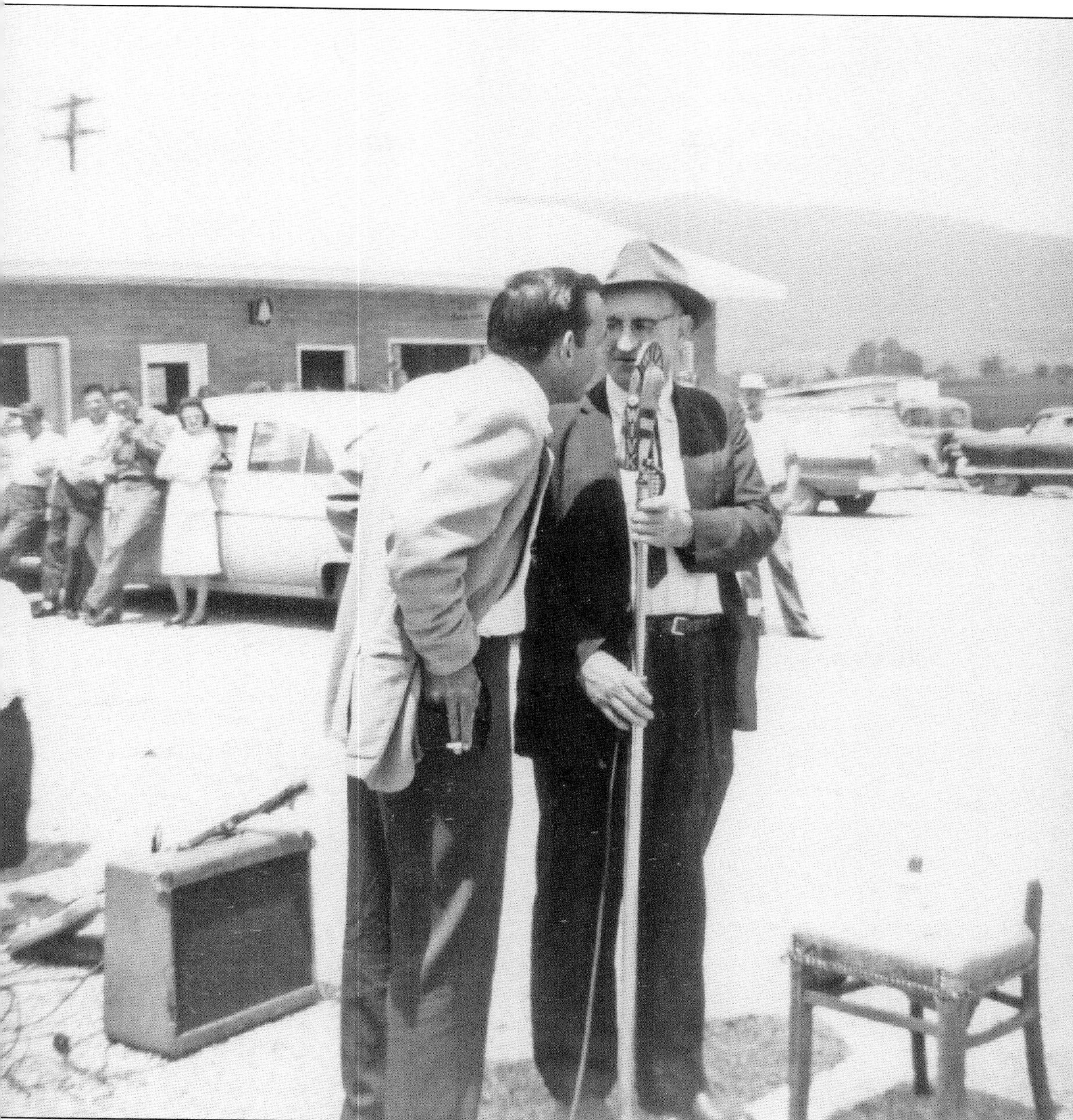

Cas Walker (right) and WIVK morning announcer Claude Tomlinson do a remote broadcast. The station was one of the local pioneers in broadcasting programs from places outside the studio that would attract listeners. (Photograph courtesy McClung Historical Collection, Knox County Public Library System, Knoxville, Tennessee.)

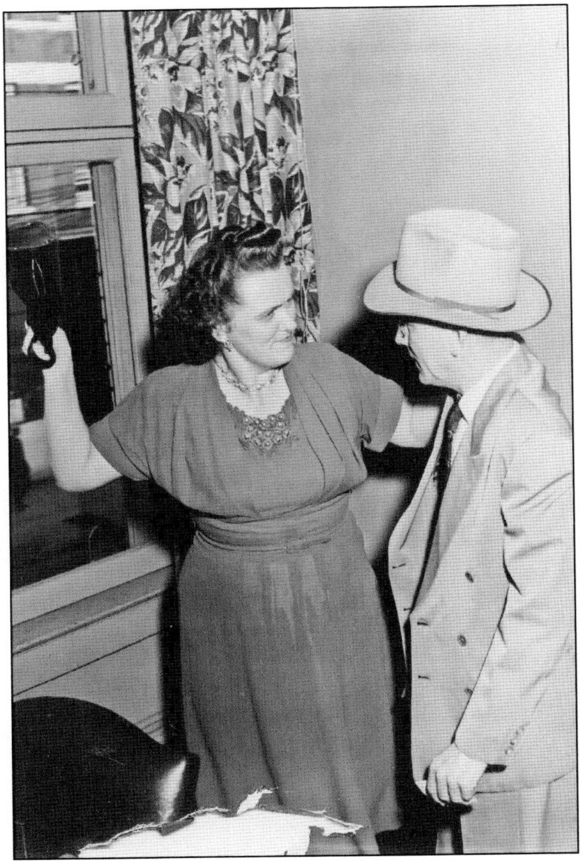

Another local icon in Knoxville broadcasting was the Reverend J. B. Mull, who hosted a program on gospel music. In spite of his fiery sermons, he was an easygoing man. Here he is pictured with his wife, who was known to listeners and then viewers only as Mrs. Mull. Below, the couple faces forward in a trademark photograph. (Photograph courtesy McClung Historical Collection, Knox County Public Library System, Knoxville, Tennessee.)

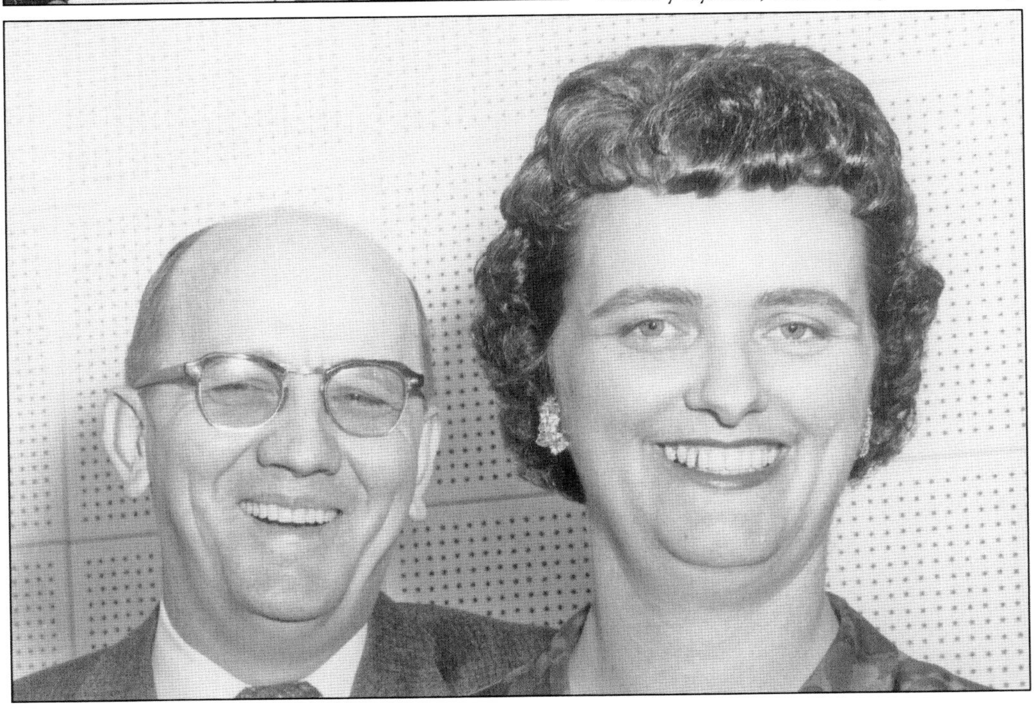

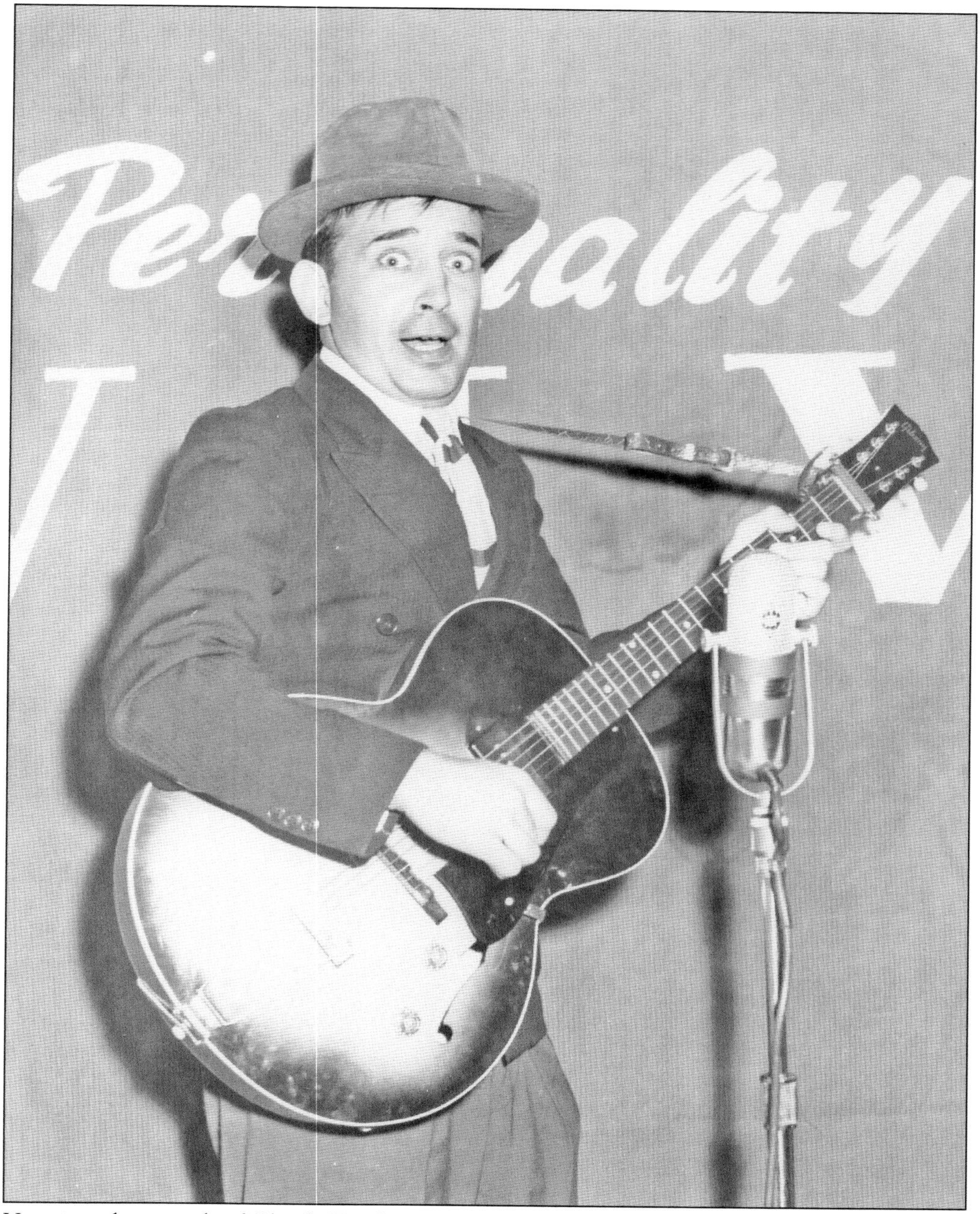

Here is a photograph of Claude Tomlinson as "Little Alf" with his guitar at a WIVK program. As Claude Tomlinson evolved as a morning announcer, he also became a well-known live entertainer. (Photograph courtesy McClung Historical Collection, Knox County Public Library System, Knoxville, Tennessee.)

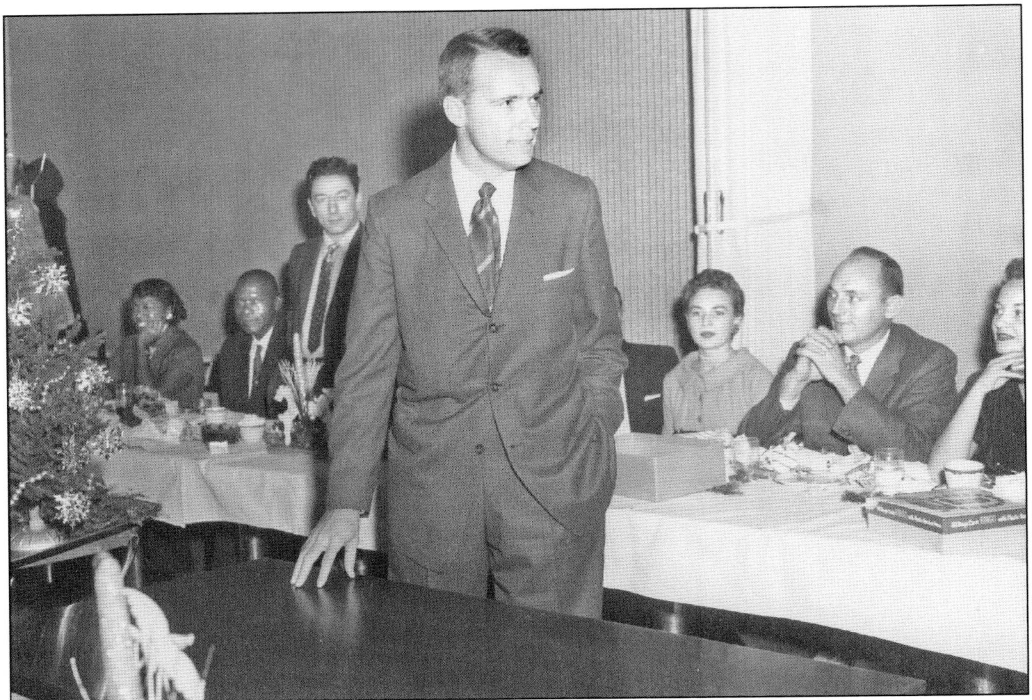

Jim Dick (standing) is pictured at a station staff Christmas party. Dick always believed in taking care of his employees and had fun with them. He is seen below with an employee who had won cash in one of the party games. He was fond of games that allowed employees to win money. (Photographs courtesy McClung Historical Collection, Knox County Public Library System, Knoxville, Tennessee.)

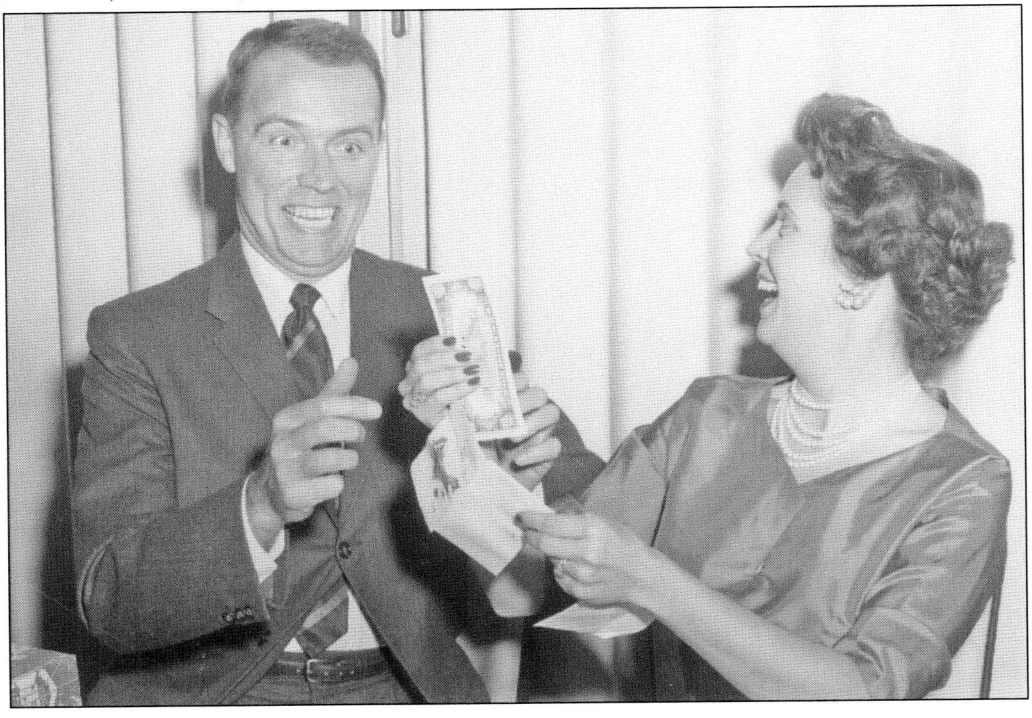

Two

THE INDEPENDENT VOICES OF KNOXVILLE

The history of those independent voices that rolled over the airwaves at WIVK reads like a who's who in American broadcasting. None, however, can replace the founding voice of morning announcer Claude Tomlinson. He was the benchmark that others tried to emulate and set a standard of excellence still talked about to this day. Tomlinson got turned onto radio when he was six years old and his father took him to the *Tennessee Barn Dance* at WNOX in Knoxville. The *Barn Dance* was one of the most popular radio programs in the South, and it made quite an impression on the young Tomlinson, who knew then he wanted to be a radio announcer. He was the oldest of four children who grew up in the city. His father worked constructions jobs, and his mother was a teacher. Tomlinson attended Young High School and participated in the school's choral department, singing in the quartet. Three months after graduating from high school, he found himself in the U.S. Army being trained as a tank radio repairman. He did one term in the service and had returned to Knoxville to go to the University of Tennessee when he went to work for Jim Dick at WIVK.

Tomlinson started small and gradually developed into a first-rate broadcaster and infamous practical joker, often engaging in a game of one-upmanship between the announcers that led to some rather interesting situations. Tomlinson would stay long after hours and work on his impressions and voices with Jim Hess, putting in the time needed to produce a first-rate morning show. In a business known for trading jobs, Tomlinson stayed put and evolved with WIVK. In 1959, the station moved from its home on Gay Street downtown to a new building on Bearden Hill, a place ironically known in the song and film *Thunder Road*—and maybe that piece of local legend played a part in reminding the station of its roots.

Tomlinson played country, jazz, and gospel, and when the station went Top 40 in 1961, so did "Claude the Cat," who simply transferred the character over to rock and roll. The format change was not so good for everyone else, and the station's leading voices like Cas Walker, Reverend Mull, A. C. Wilson (Knoxville's first black disc jockey), and popular female broadcaster Laurie June (Knoxville's first female broadcaster) all left the station. Archie Campbell, who had also been an early part of the station's broadcasts, had moved on to Nashville. It was a three-year ride as a Top 40 station before WIVK returned somewhat to its original format. During that time, Tomlinson also acted as a program director of sorts and heard a young man he described as having a deep steady voice at a competing station; Claude had come to recognize talent, so he hired a young Bobby Denton to take over the afternoon shift at WIVK.

Bobby Denton became a broadcast icon himself in the city. The former Maryville Drag Strip announcer also made his home at the station and rose through the ranks to become general manager. WIVK was never quite the old station after that. Nashville was producing a new sound of country music that made the transition from Top 40 a little less shocking and brought forth a new sound that became its trademark. WIVK never advertised itself as the country music station but as the "Music of America" station instead.

A year after the changeover, something happened at WIVK that would change the station forever. In September 1965, according to Claude Tomlinson, station staff watched the clock, waiting forever before news finally came that the FCC had given WIVK AM 850 a license to operate as WIVK FM 107.7. No longer was a broadcast day from sunup to sundown—they now could operate 24 hours a day, seven days a week. There were no digital radios in those days, so the station, for promotion purposes, always advertised itself at 108 FM.

While AM was still king of the dial, Jim Dick, ever the promoter, purchased thousands of FM converters and started the "half a radio" giveaway promotion, which encouraged people to come by the station and get a free converter for their car and make it a "whole radio." The promotion worked brilliantly, and WIVK FM soon became a powerhouse station in Knoxville.

The Music of America had become a popular format for the WIVK listeners. The late 1960s also saw the station begin to take shape into its present form. Kenny Dearstone and Norman Majors—who was WIVK's first named news director—did mornings; Claude Tomlinson did 10:00 a.m. to noon; a new voice named Ed Brantley, who joined the station in 1968, did 1:00 p.m. to 3:00 p.m.; Bobby Denton, 3:00 p.m. to 7:00 p.m. ; Mike Hanes, 7:00 p.m. to midnight; followed by the "Midnight Rider," Terry Womack, from midnight to 6:00 a.m.

The AM side of the station covered its bases as well. WIVK AM went from 1,000 watts to 50,000, moved from 860 to 850 on the dial, and became known as the "Jolly Green Giant" station. With both AM and FM frequencies pumping out night and day, the station was reaching more homes, cars, and businesses than ever before. The backbone of radio—advertising sales—suddenly started booming for the station, which created new jobs and opportunities for those independent voices to be heard.

In 1966, with no more station auditorium for listeners, WIVK still brought the nation's best country music talents to Knoxville, with its *Shindig* concert series debuting a new singer named Kenny Rogers. As the 1960s came to a close, Claude Tomlinson became general manager of WIVK AM and FM, but it was not a job he really wanted, so he returned in 1972 with the *Great Day Show with Claude, Lester Longmire, and Old Man Schultz*. Broadcaster Dave Young, who came up with the "wivick, wivick" that made a frog one of the most recognized station mascots in radio, joined the station in the 1970s, as did Jimmy Vineyard, whose Sunday night broadcasts on the latest in bluegrass music gave rise to a new field on the Music of America station.

Competitors often lamented that listeners would buy radios, tune it to WIVK, and break the dial. WIVK ratings were unbeatable. The anchor announcers stayed the same, which speaks well of Jim Dick and the feeling of family that pervaded the station. It is unheard of in this business, but it was a formula that could not be beat, and East Tennessee never stopped listening. As country music rose in popularity, WIVK was no longer the new kid on the block. It found itself becoming a heritage radio station. Here are the voices and the sounds that made WIVK what it is today.

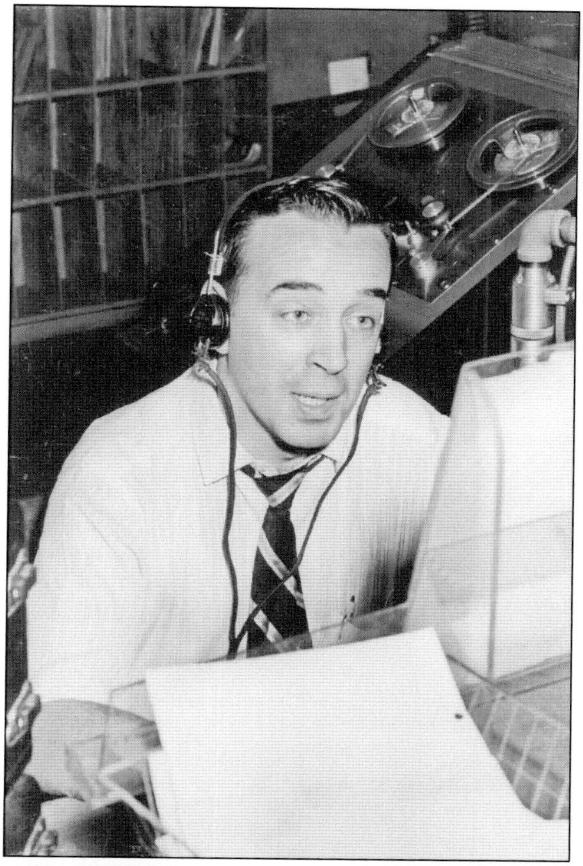

Claude Tomlinson was a man of many voices who worked late in the evenings after the AM had signed off of the air to perfect his various characters. Broadcasters marveled at how he could turn them on and off without thinking. Here Tomlinson dresses the part of a respectable businessman for station promotion shots. The characters never changed, though, especially as the station moved out of block programming and into what is known today as modern radio. It was the constant that kept listeners tuning in his shows.

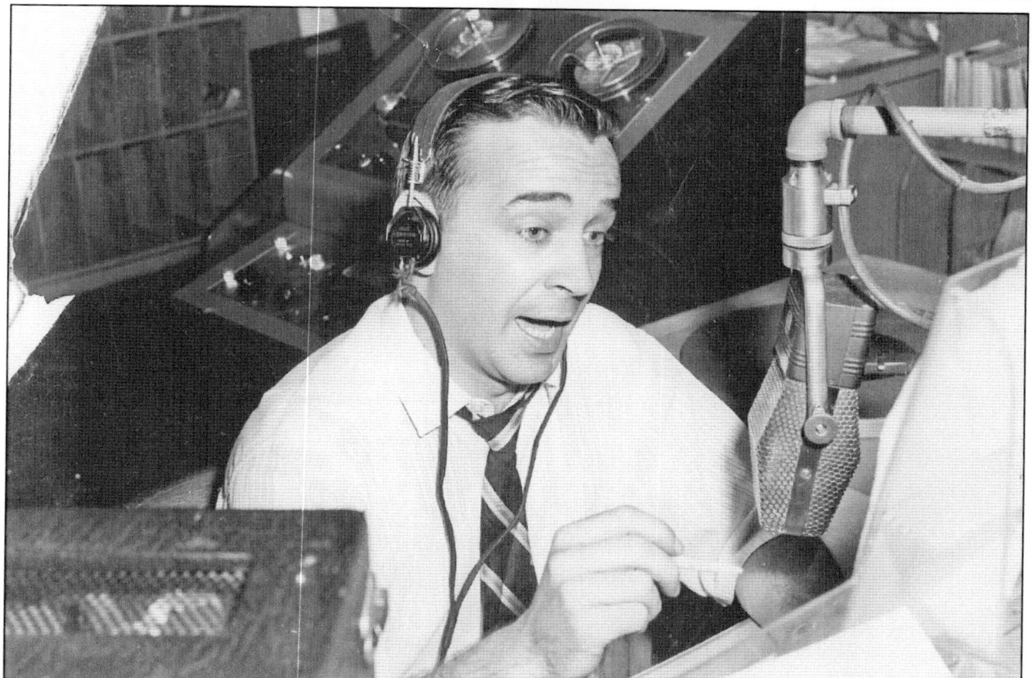

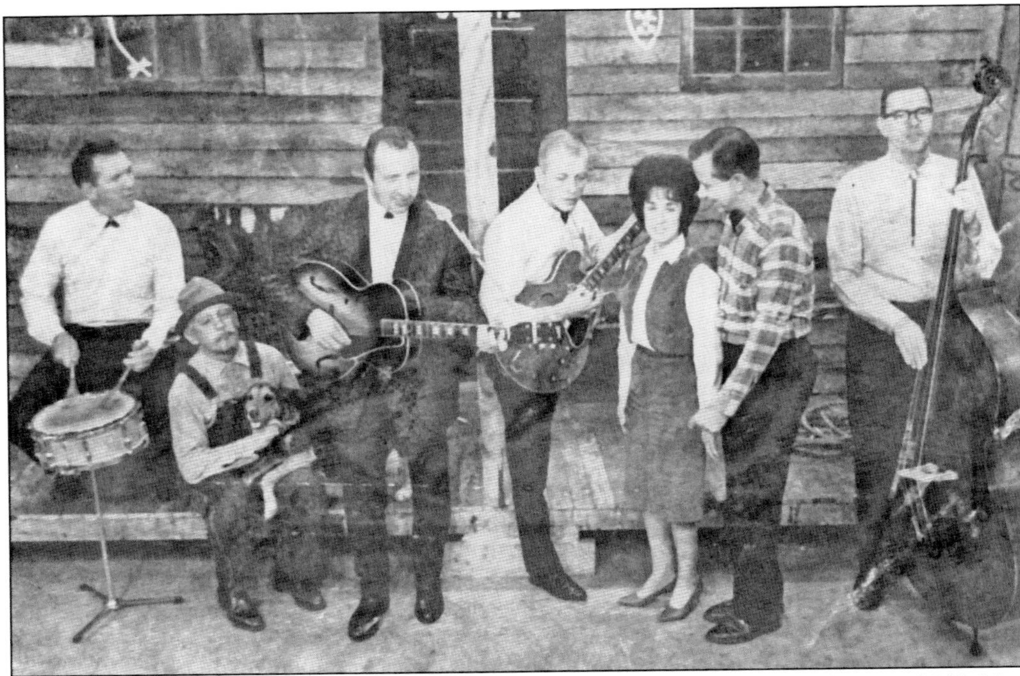

Art Lavidge of Lavidge and Associates Advertising decided to take Tomlinson's "Old Man Schultz" character to television when he wanted to do a country music variety hour. Once a week, Tomlinson dressed the part and introduced Knoxville to new country talent. "Old Man Schultz," second from left with his dog Buford, listens to the *Shultz Show* regulars, which included Lynn Clapp, who would go on to found Lynn's Guitars, today's Broadway Sound. Below is a billboard for the show.

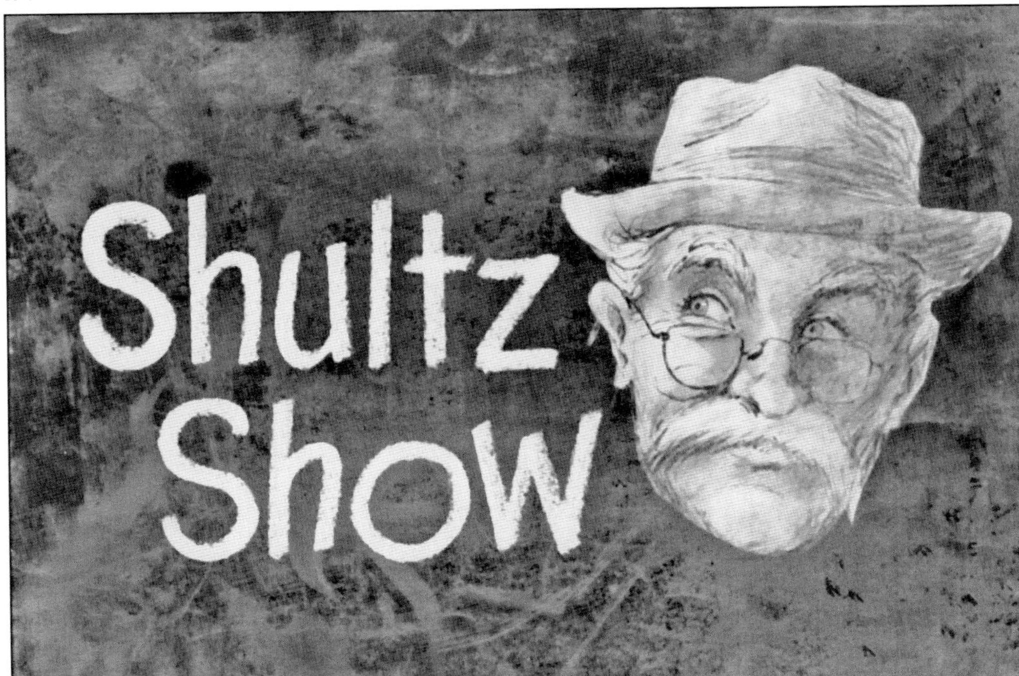

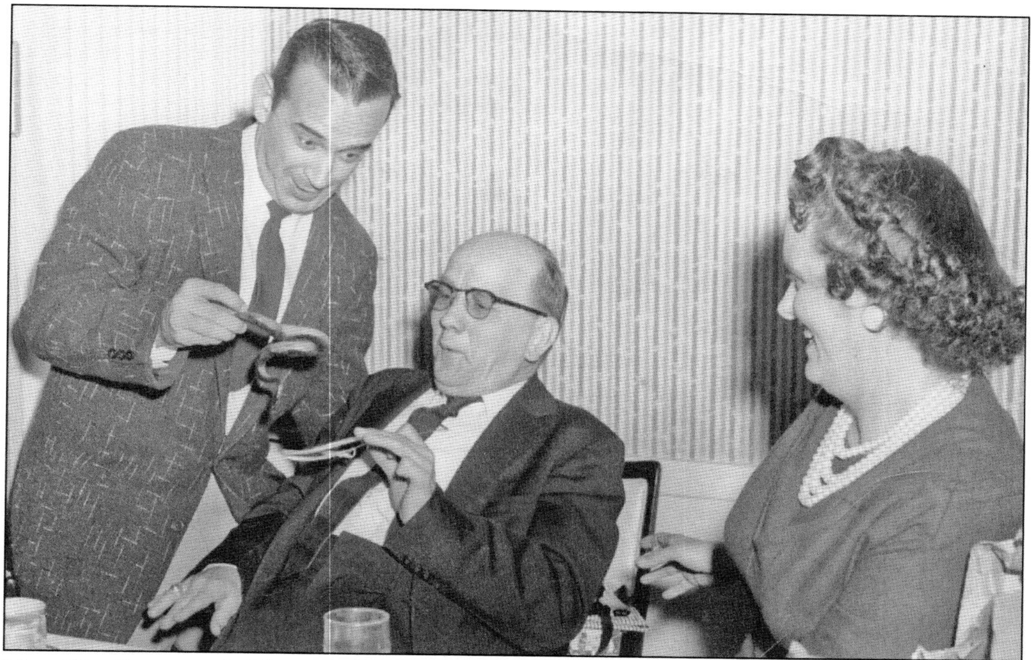

Here Tomlinson is holding a plastic snake that has the Reverend J. B. Mull climbing out of his chair. The announcer was a prolific practical joker and was known for his pranks on other announcers.

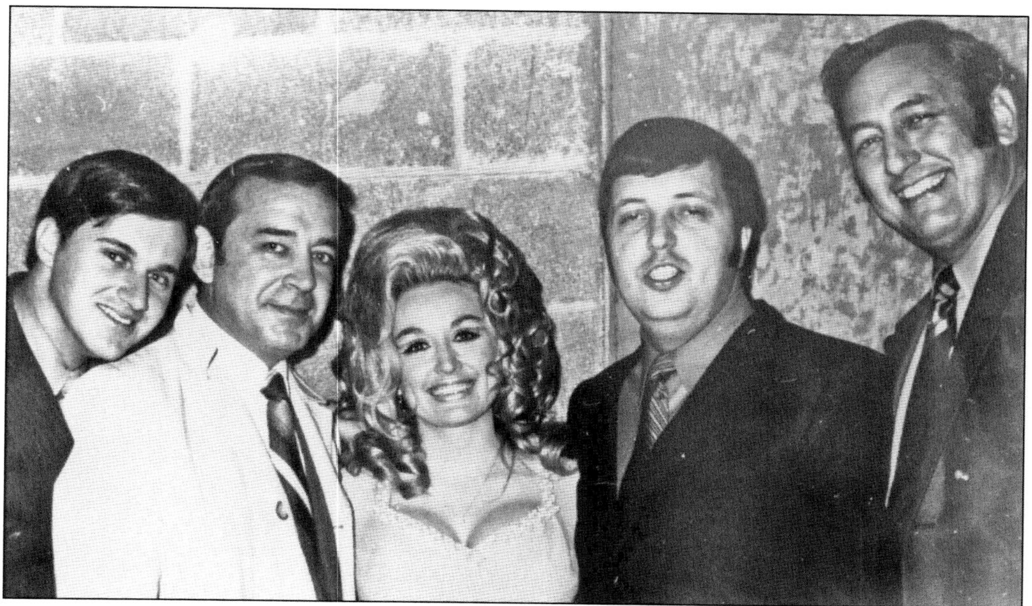

Tomlinson supplemented his income at the radio station by doing some production work on the side. In the 1960s, he began recording some songs with a family in Sevierville. In 1961, he engineered a song titled "So Little I Wanted, So Little I Got" at WIVK's studios. It went on to become a popular duet and helped launch the career of country music entertainer Dolly Parton. The Partons always remained close with the station. Here she is pictured with Bobby Denton (second from right), Ed Brantley (far left), and record executives. (Photograph courtesy WIVK.)

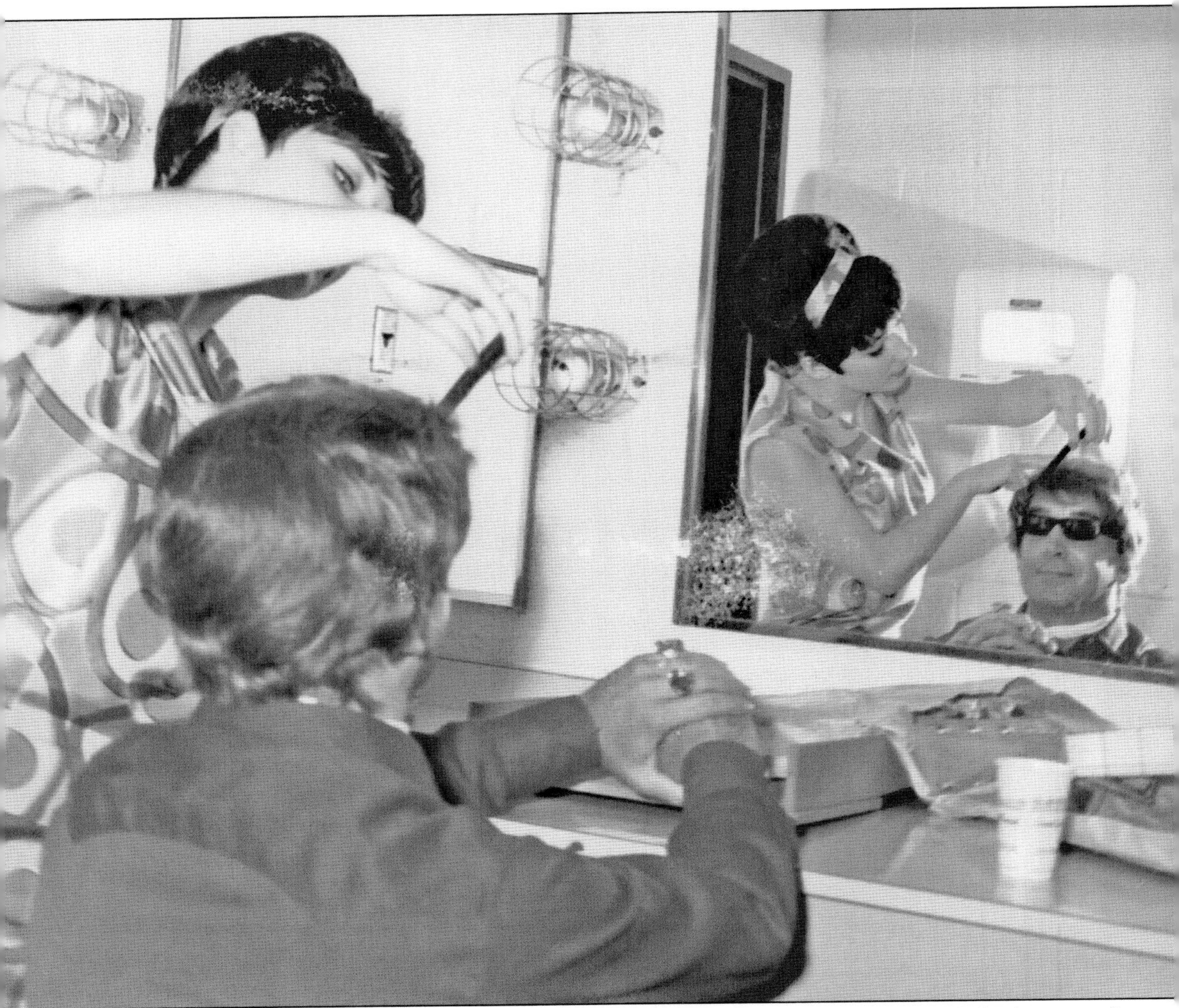

Tomlinson still could not get the onstage performance bug out of his system. With the rise of Hank Snow, he developed the character "Hank Range." Here he is in make-up getting ready for a stage performance.

Here Hank Range is seen on stage with Kenny Dearstone, who was an announcer at the station. Tomlinson played the role so well, many people never knew who he was.

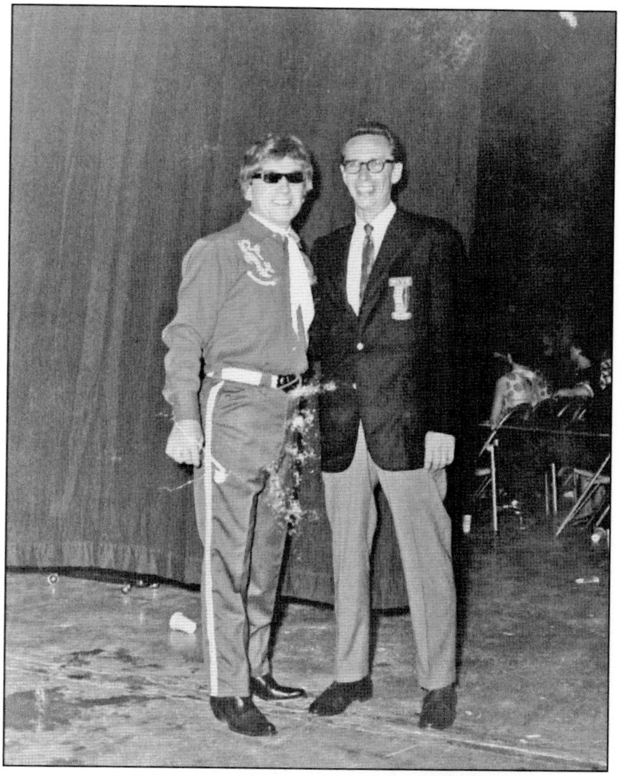

Station parties and get-togethers were always special occasions. From left to right are Ken Russell, Claude Tomlinson, and Bobby Denton at a WIVK Christmas party toasting to friendship and a good year for the station.

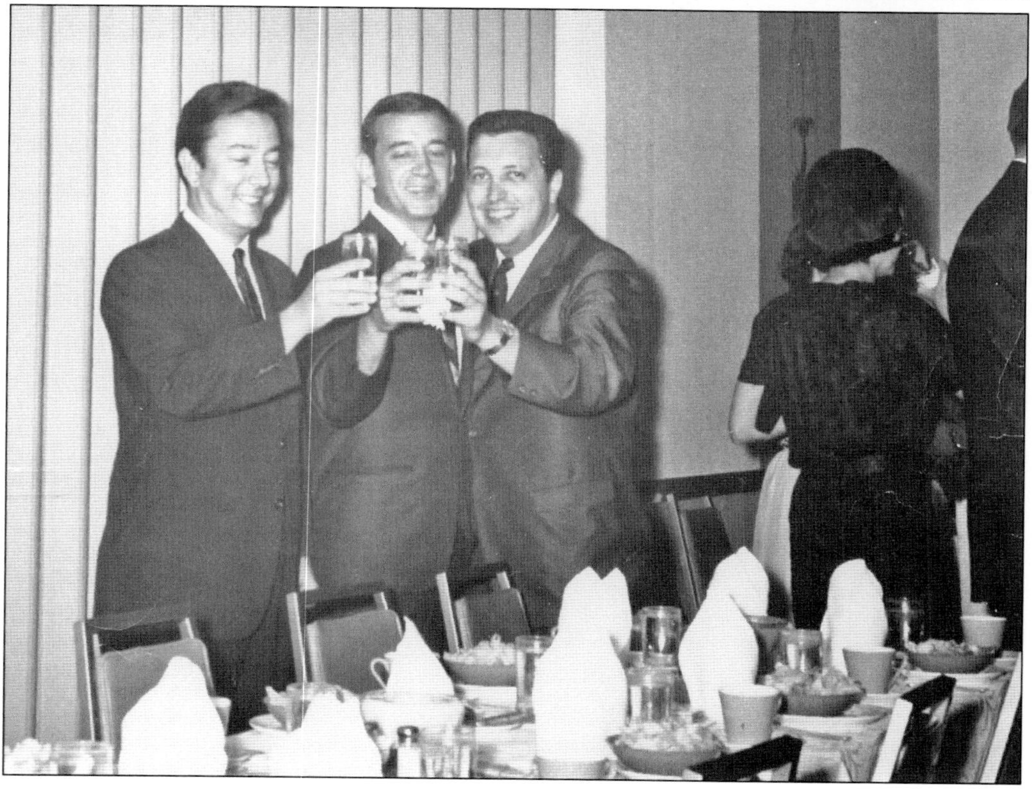

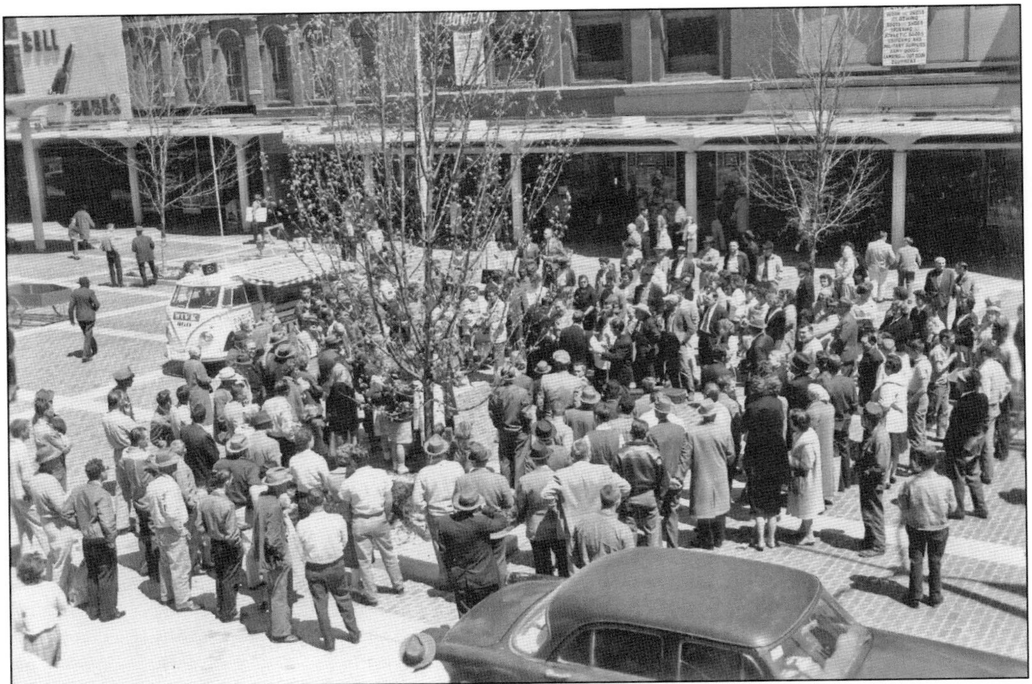

WIVK was one of the first stations to see the value of doing live remotes. Here the WIVK van is parked on Market Square downtown. The crowd would always gather to see a radio broadcast in person.

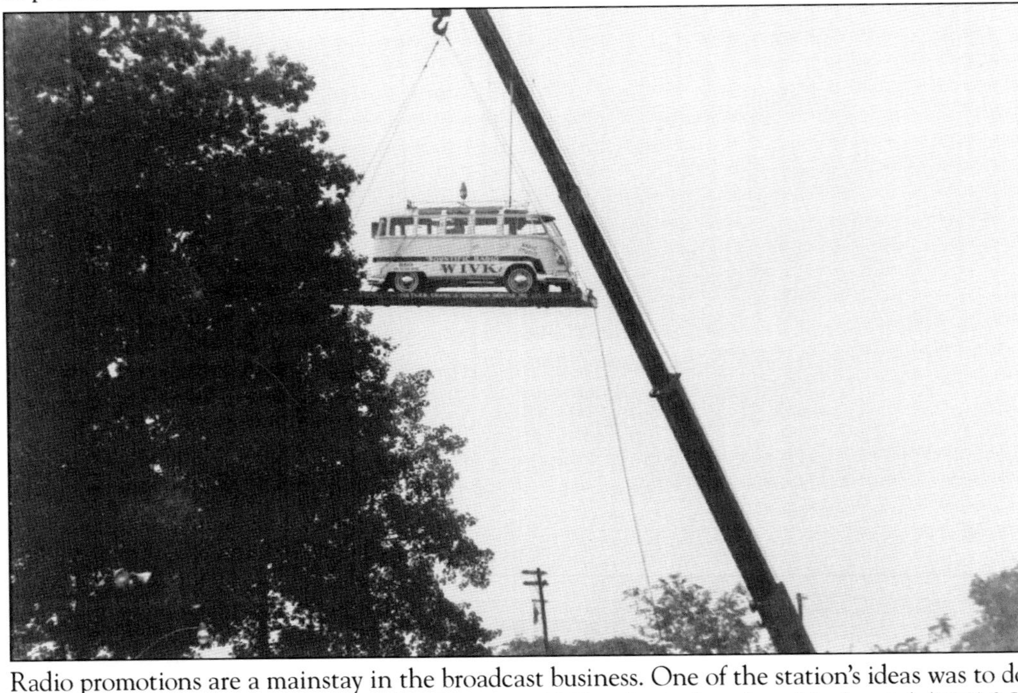

Radio promotions are a mainstay in the broadcast business. One of the station's ideas was to do an "aerial" broadcast from the yearly Tennessee Valley Agricultural and Industrial (TVA&I) Fair. The station van was loaded onto a crane and suspended so Tomlinson could do a broadcast overlooking Chilhowee Park fairgrounds.

After the van was suspended, Claude Tomlinson climbed up and did his morning show from the perched vehicle. It was no small feat for the radio announcer, as Claude Tomlinson had a major fear of heights. In the photograph below, it is hard to see, but Claude Tomlinson is waving to listeners from the window of the WIVK van. It was a remote broadcast the morning man would never forget.

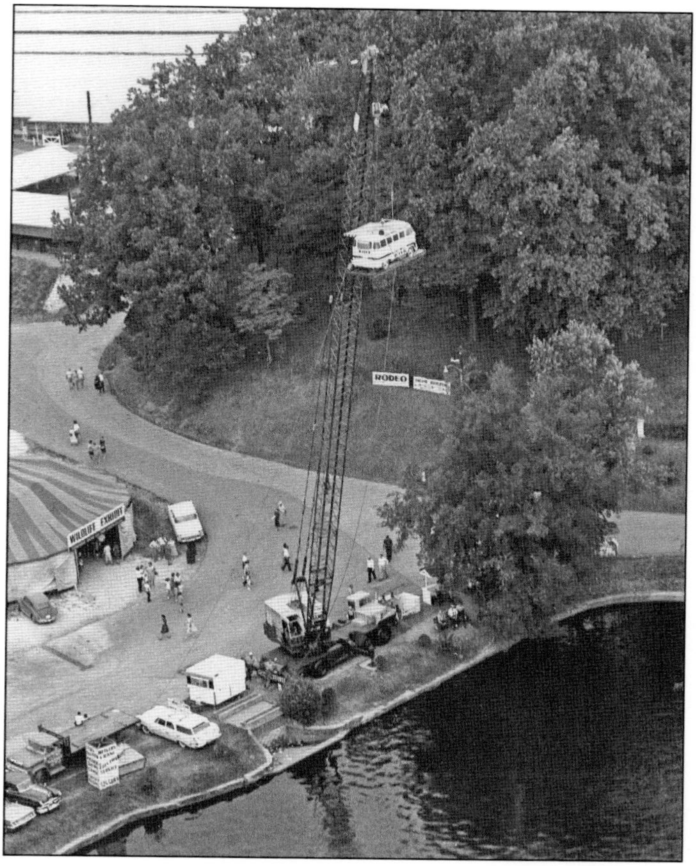

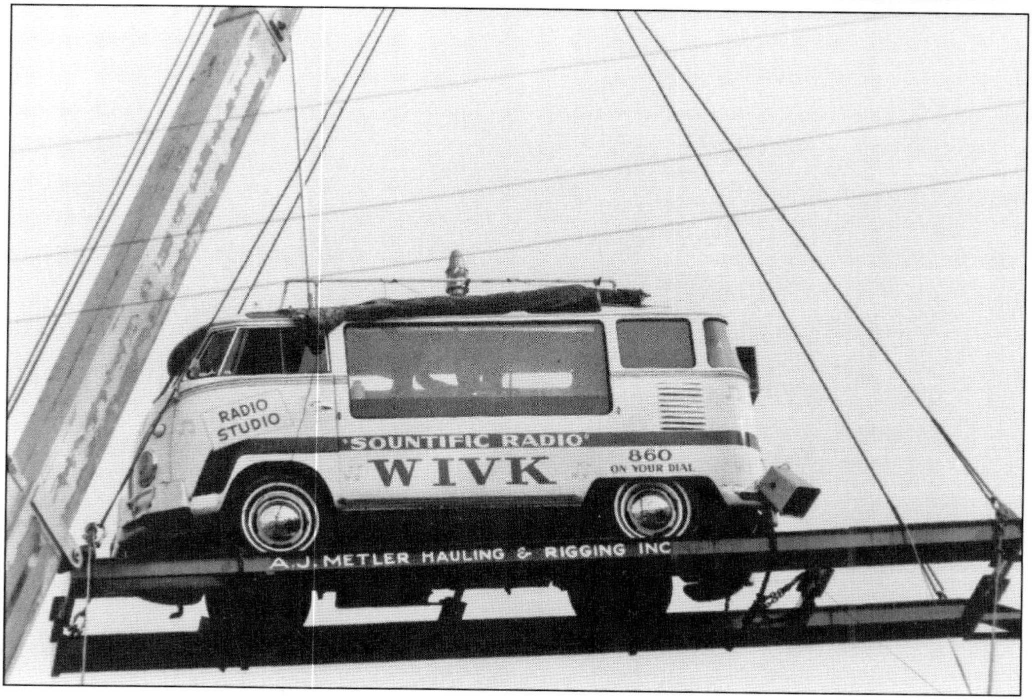

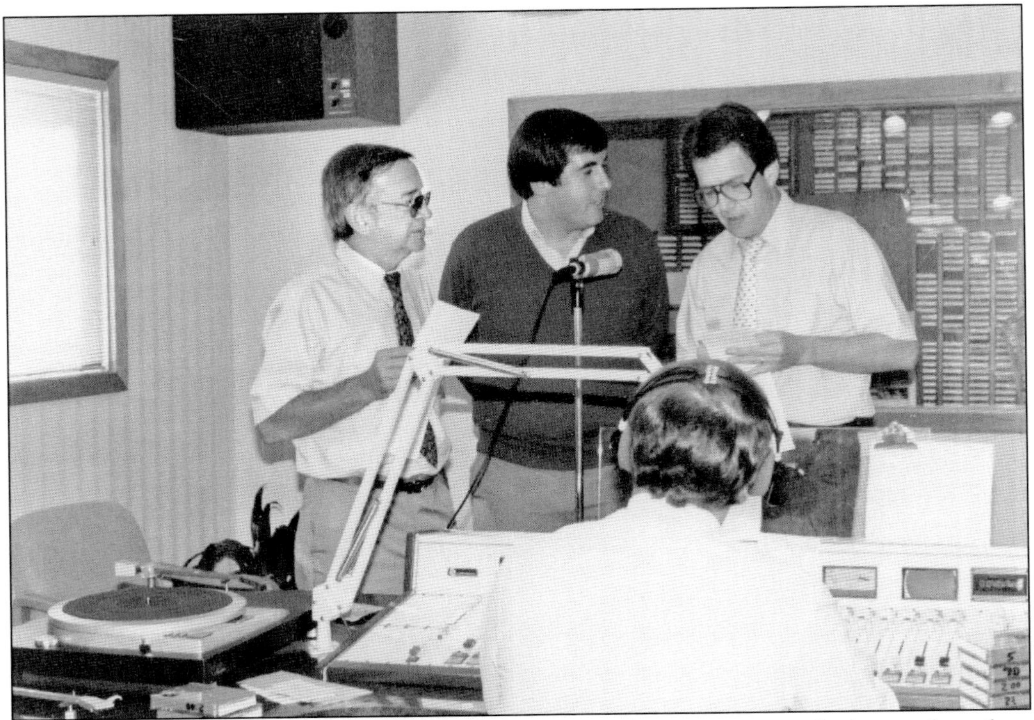

The morning crew had a number of members over the years that went on to other stations. Standing from left to right are Claude Tomlinson; current Voice of the UT Volunteers Bob Kesling; and longtime Knoxville broadcaster Curtis Parham, who once hosted a morning show on WIVK AM with broadcaster Walker Johnson. (Photograph courtesy WIVK.)

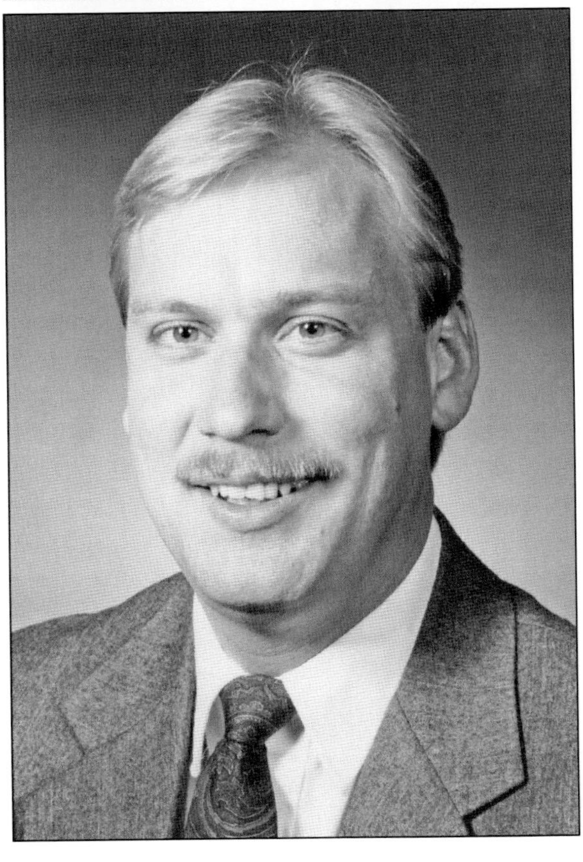

One of the most beloved members of the WIVK family was announcer and production director Jim Donovan, who passed away from cancer. He was a vibrant part of the station's on-air sound and a driving force behind the success of the *Great Day in the Morning* show. (Photograph courtesy WIVK.)

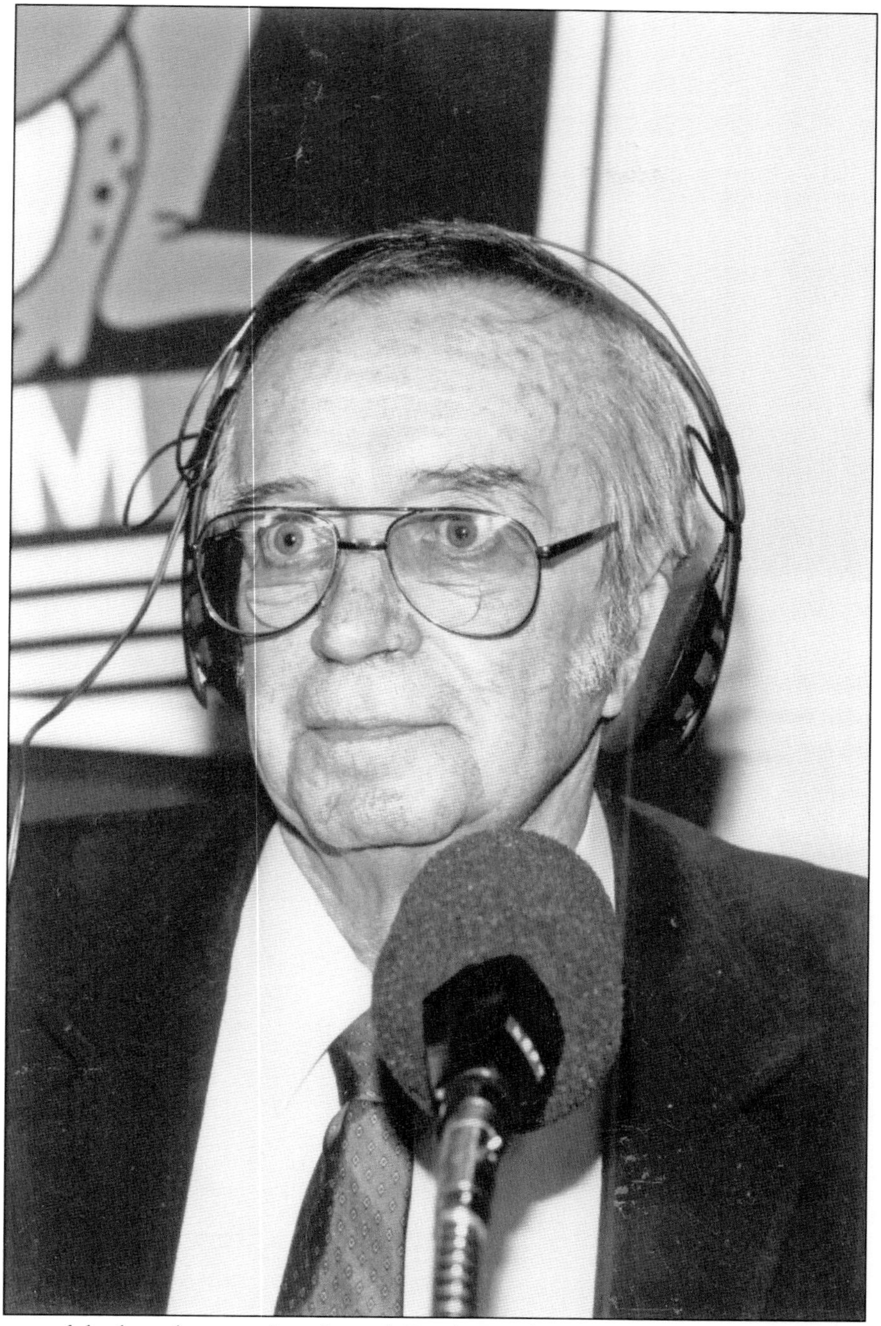

This is one of the last photographs taken of Claude Tomlinson before he retired in 1991, almost 30 years to the day after he first signed the station on the air. He died two months later, and a remarkable chapter in East Tennessee broadcasting was closed. During his time, Tomlinson helped take the station to new heights and was nominated by *Billboard* magazine as Personality of the Year in 1988. His show was part of the reason the station was named the No. 1 radio station in America in all formats by trade publication *Radio and Records* with an incredible 46.6 rating—meaning the station had the highest percentage of listening audience in the nation. (Photograph courtesy WIVK.)

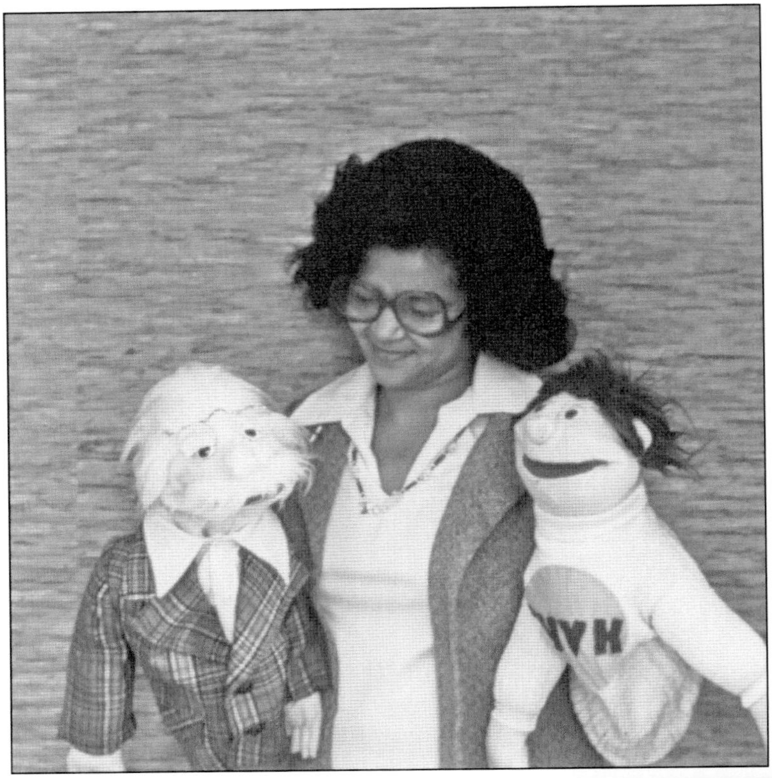

Lester Longmire and Old Man Schultz were not only radio characters. Here Ann Robinson shows the puppets that were often used in radio promotions of the *Great Day in the Morning* show.

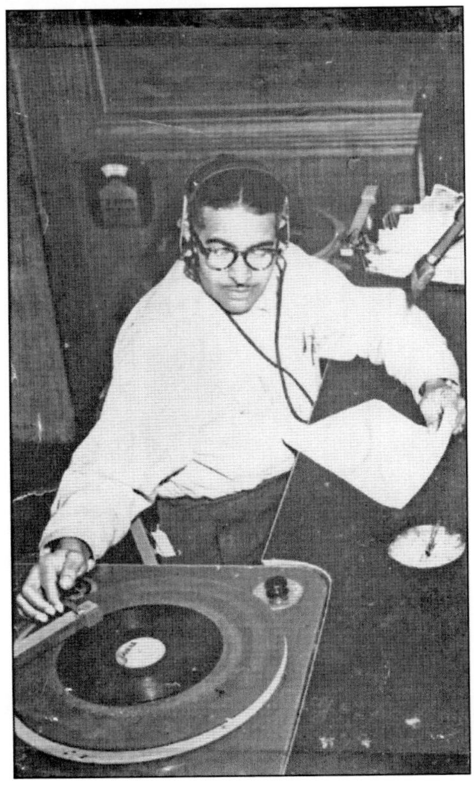

WIVK's A. C. Wilson was Knoxville's first black broadcaster and was known for his program. He and Tomlinson used to go back and forth with practical jokes with each other. He was once rated by *Color* magazine as one of the nation's most outstanding disc jockeys. He hosted an evening jazz show and later did the *Acey Boy Show*, which earned him recognition in Knoxville's black community.

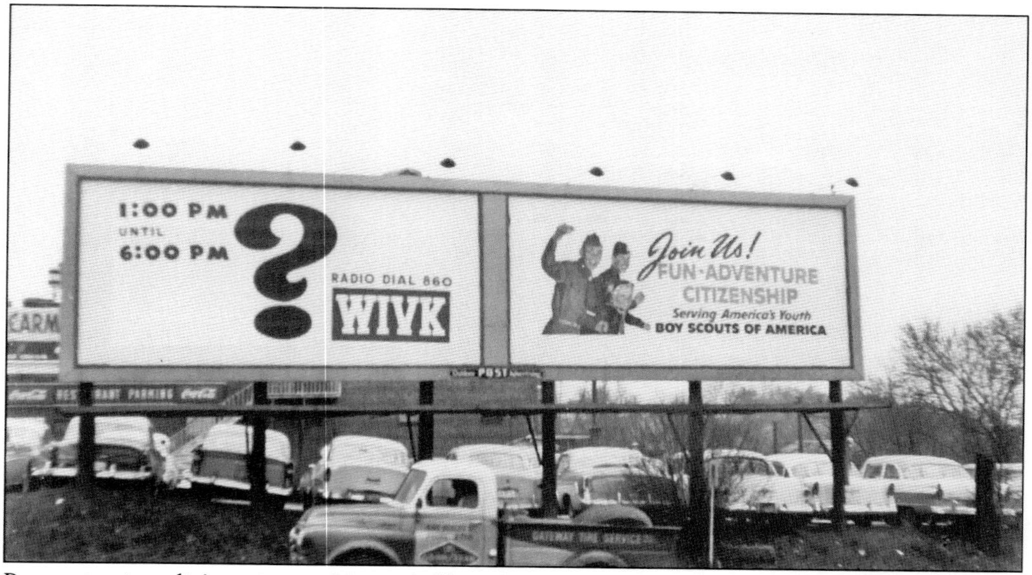

Promotion is radio's mainstay. Here a billboard in Knoxville advertises a mystery program that kept listeners tuning in to see who was going to be on the air.

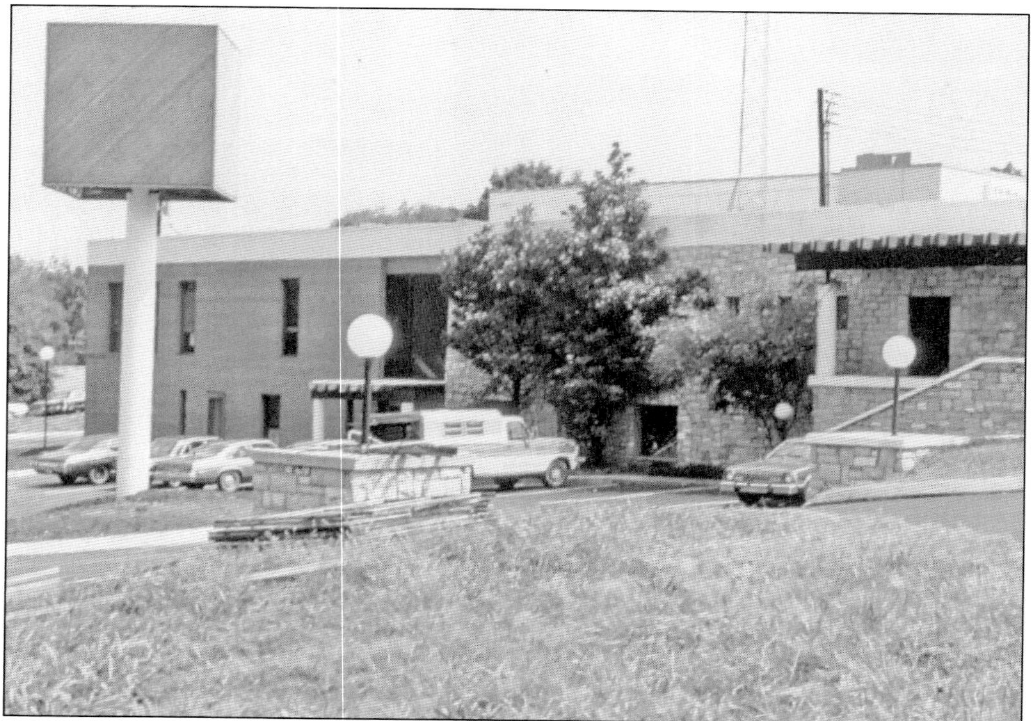

This is WIVK's second home at 6711 Kingston Pike. It was rumored to be an underground medical clinic at one time and sat on Bearden Hill.

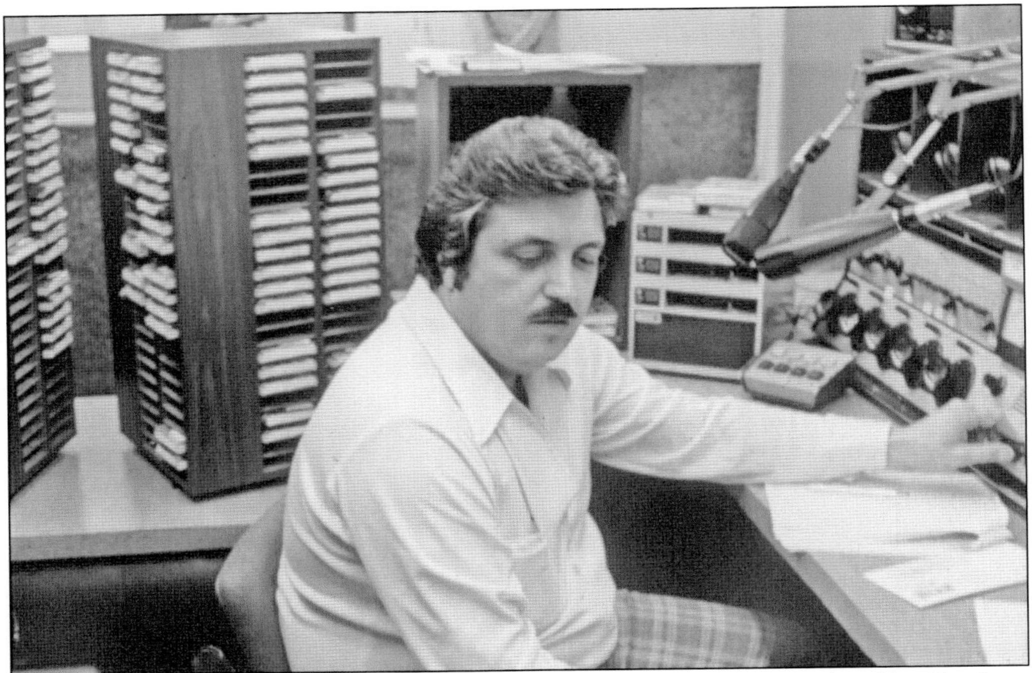

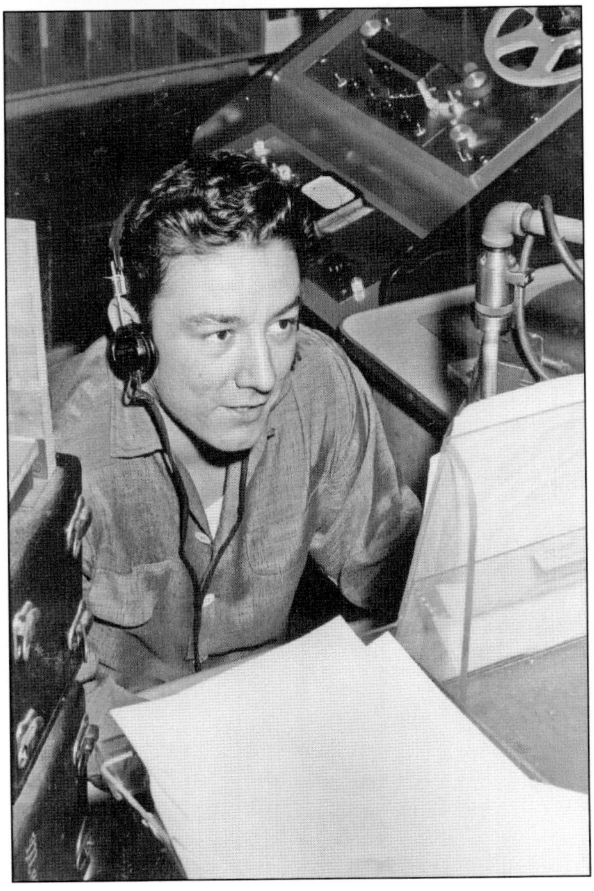

Bobby Denton was hired by Claude Tomlinson to work on-air at the station. Denton did the afternoon drive and became an icon of the station who led it to its No. 1 rating. He rose through the ranks of WIVK to become general manager. In 1992, Denton became vice president of the Country Music Association's board of directors.

Announcer Ken Russell was one of the station's most popular disc jockeys. He started at WIVK as a fill-in, but "Cousin Ken," as he was known, had an ear for local music that made his program one of the most popular on the station.

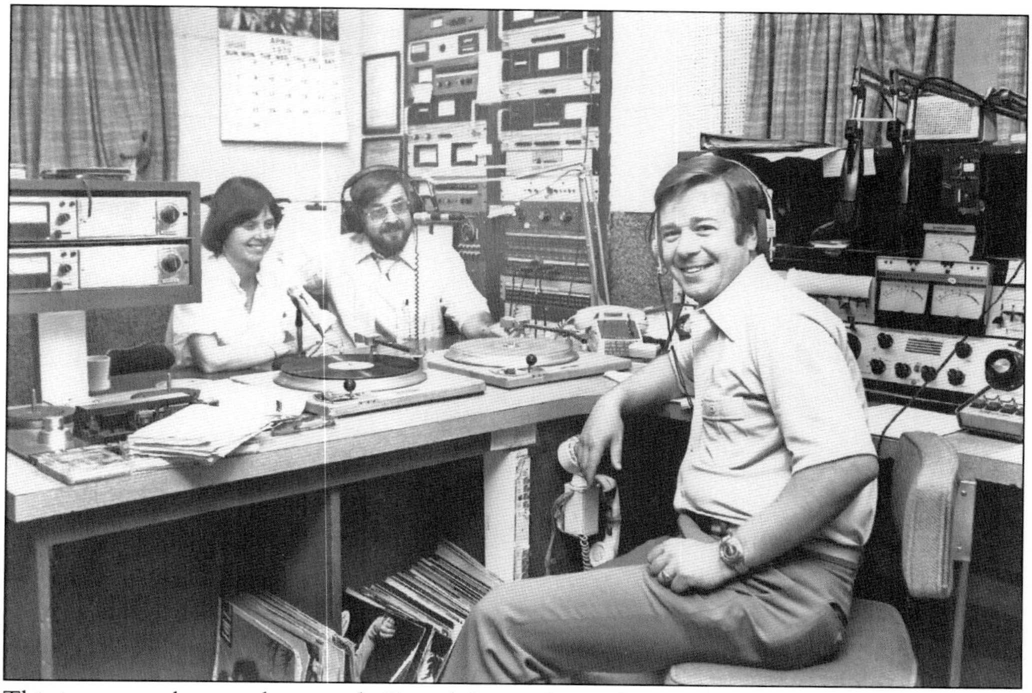

This is a control room photograph. From left to right are Jean Ash, Dan Bell, and Dave Young. "Your Dave Young," as he was known to listeners, came up with the on-air hijinks of "wivick, wivick," that led to the station adopting a frog as its mascot.

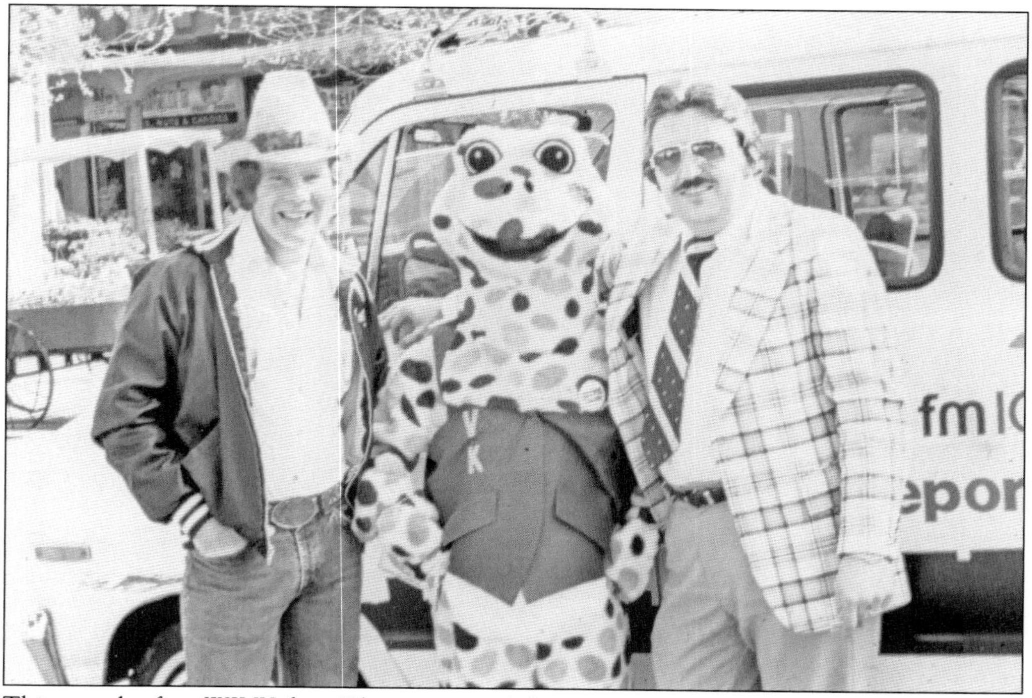

This was the first WIVK frog. The concept of the mascot came from the popular San Diego Chicken of the 1970s. Many people have played the frog over the years. Here he is pictured with Bobby Denton (right) in front of the station van.

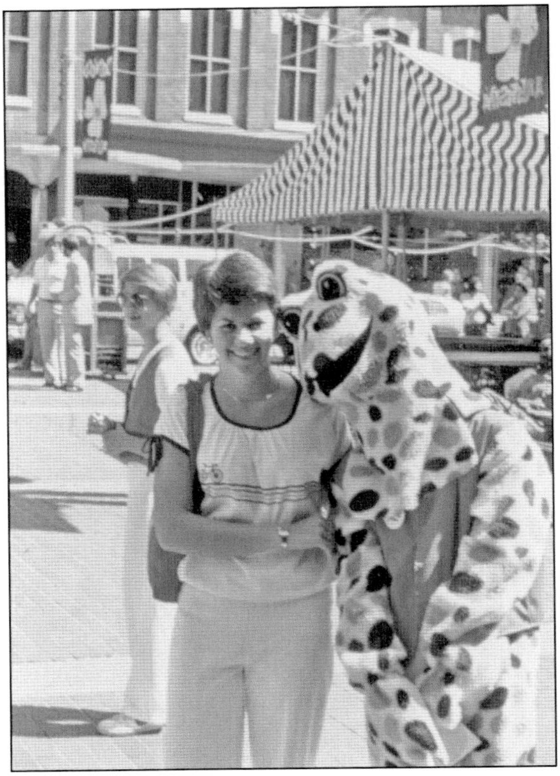

Like most station mascots, Wivick the Frog attends remotes, parties, and charity events for the station. The tan and multi-colored spotted outfit was soon changed to the WIVK green known today.

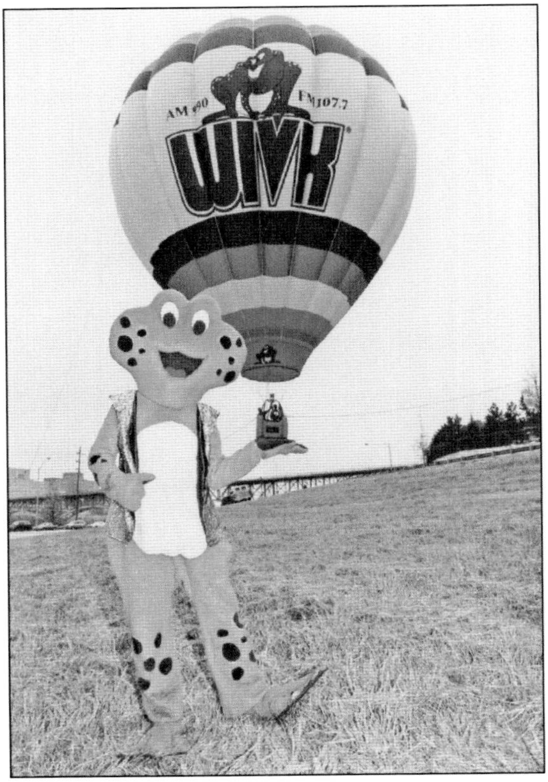

This is the modern Wivick the Frog. Although only known for its trademark "wivick, wivick," its one of the independent voices that became synonymous with the station.

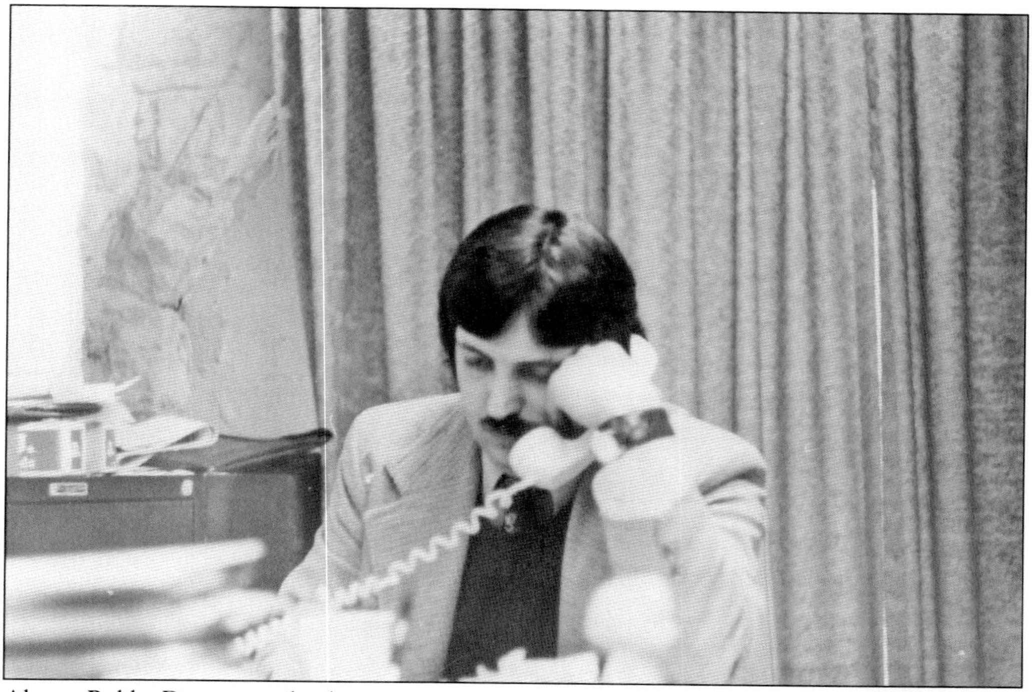

Above, Bobby Denton works the telephone. He was one of the rare people who moved on to programming, sales, and management, and he made his mark in each one. His voice is still heard on football Saturdays at Neyland Stadium in Knoxville.

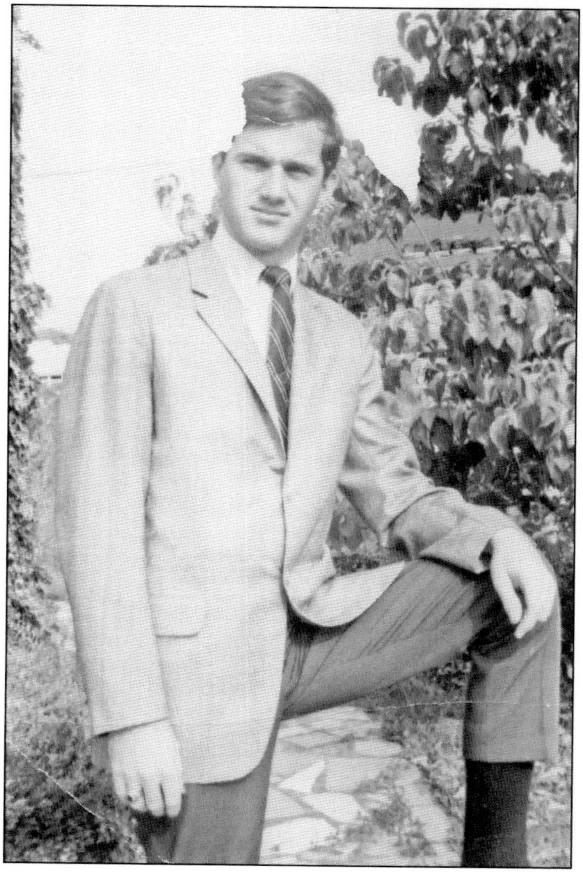

In 1968, Ed Brantley joined the ranks of WIVK. Here is a station publicity still of him. Brantley also was an announcer who worked himself up the ladder of success at the station. (Photograph courtesy Ed Brantley.)

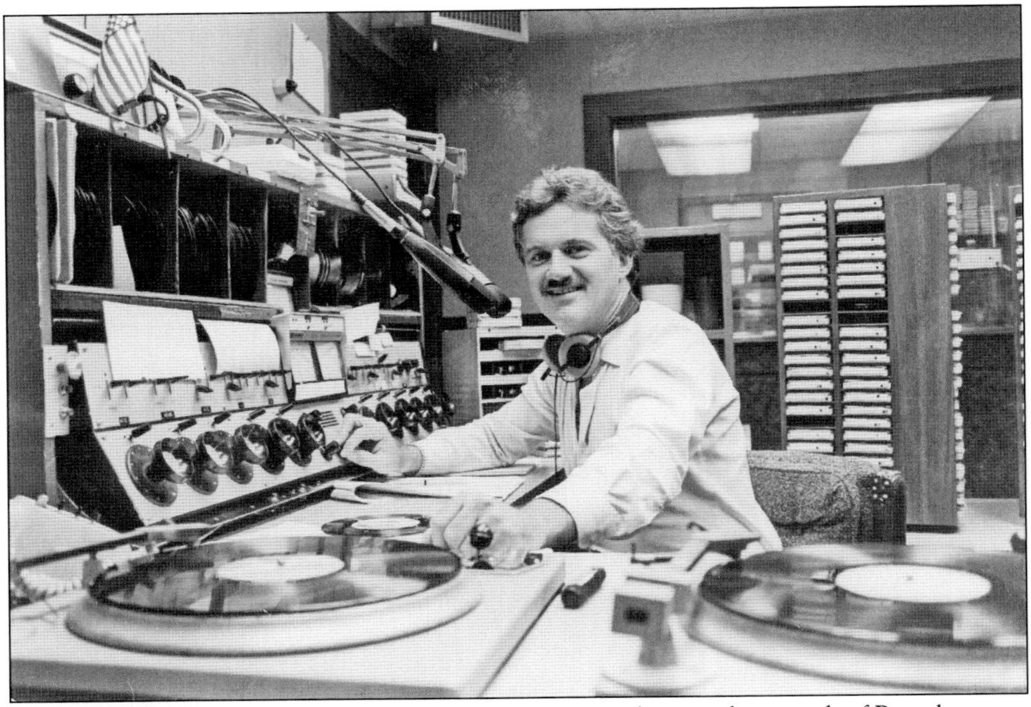

This is a photograph of Brantley in the control room of the station. He was eventually moved to the afternoon drive and was Knoxville's most listened-to disc jockey in the afternoon. His antics and following were a boom for station promotion. He laid the groundwork for the afternoon drive-time slot that others would follow. (Photograph courtesy of WIVK.)

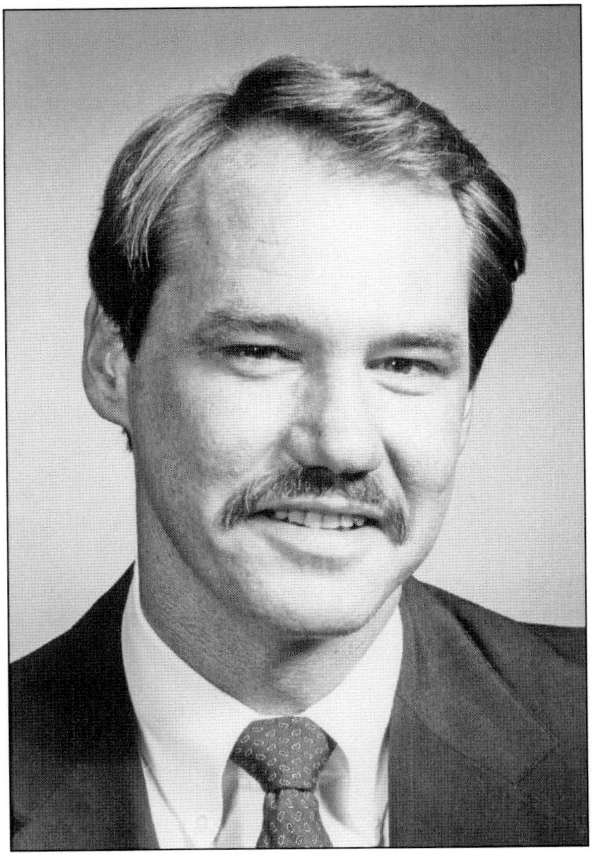

Steve Armistead, who worked on the *Great Day in the Morning* show, was also a frequent guest on the afternoon show with Ed Brantley. He was dubbed the man of many voices and said he could imitate anyone.

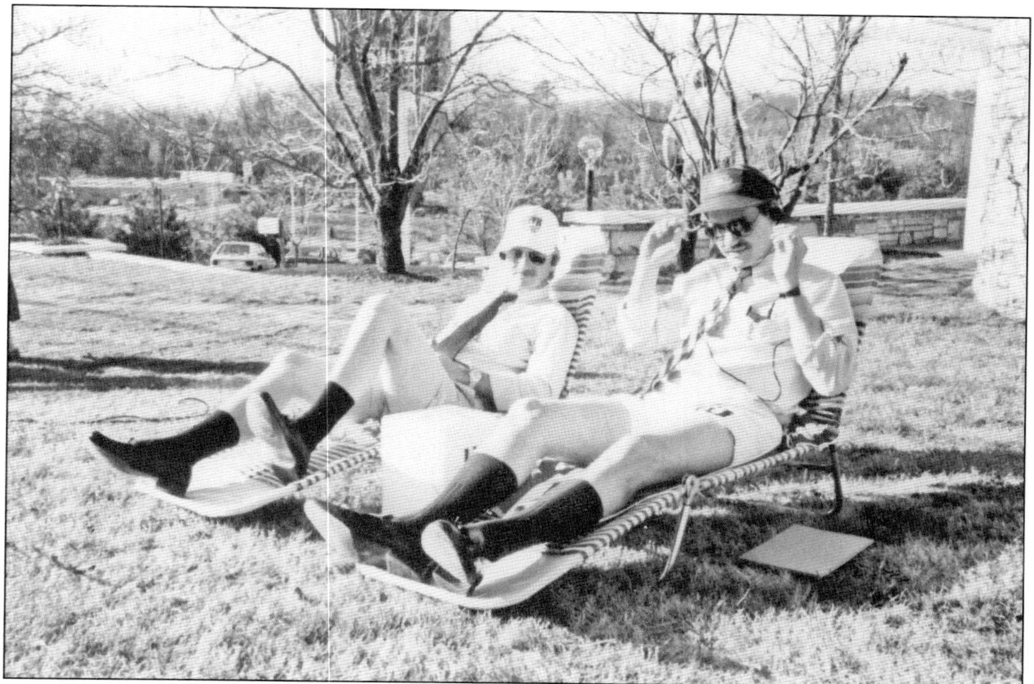

After a brutal winter in 1993 in East Tennessee, Brantley (right) and Armistead decided to do a remote outside the studio in lawn chairs on a warm spring afternoon. Chairs parked on the side of Kingston Pike made rush hour an interesting time for commuters.

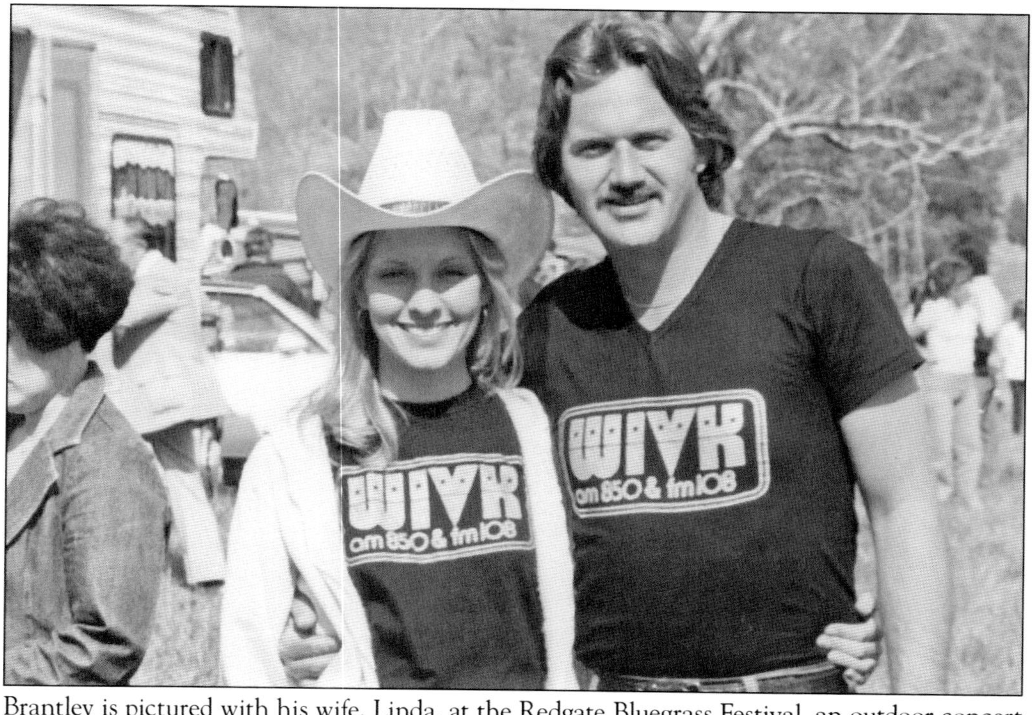

Brantley is pictured with his wife, Linda, at the Redgate Bluegrass Festival, an outdoor concert series held every year at the Redgate Farms that featured some the best national talent in bluegrass and country music. (Photograph courtesy Ed Brantley.)

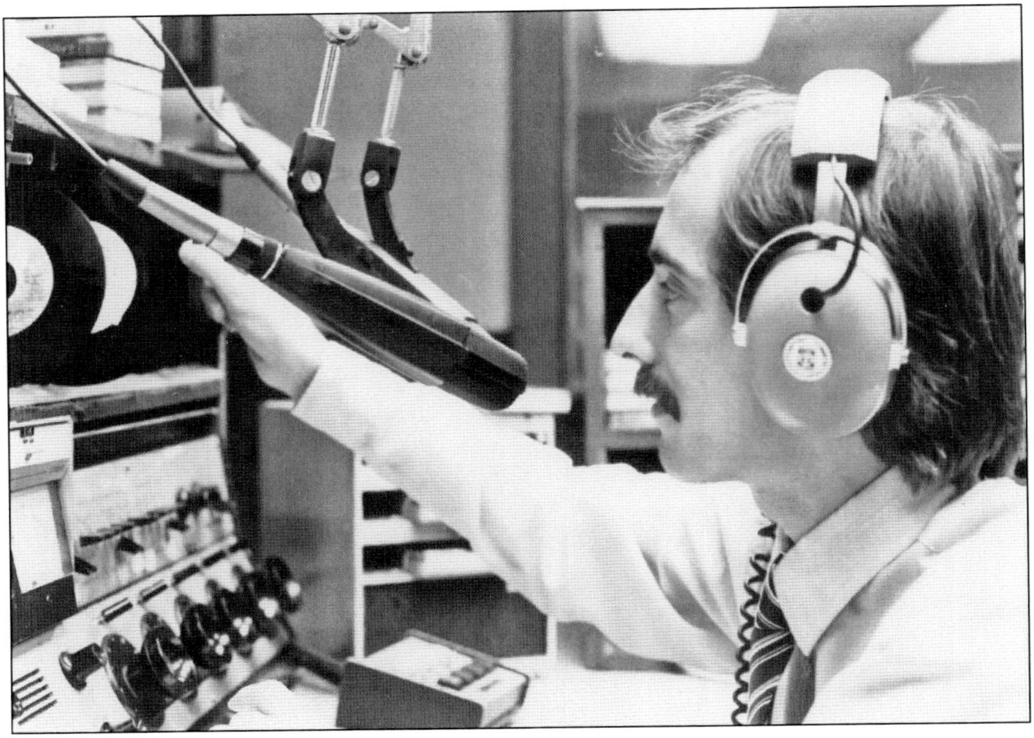

Above is an image of former afternoon announcer Bob Thomas, who went on the Hollywood to work in the broadcast industry there. Below is an image that shows it is not all talking on the air. Announcers have paperwork, too. They had to keep logs of their broadcast time as part of the FCC regulations and check off a list of commercials scheduled to be aired on the station.

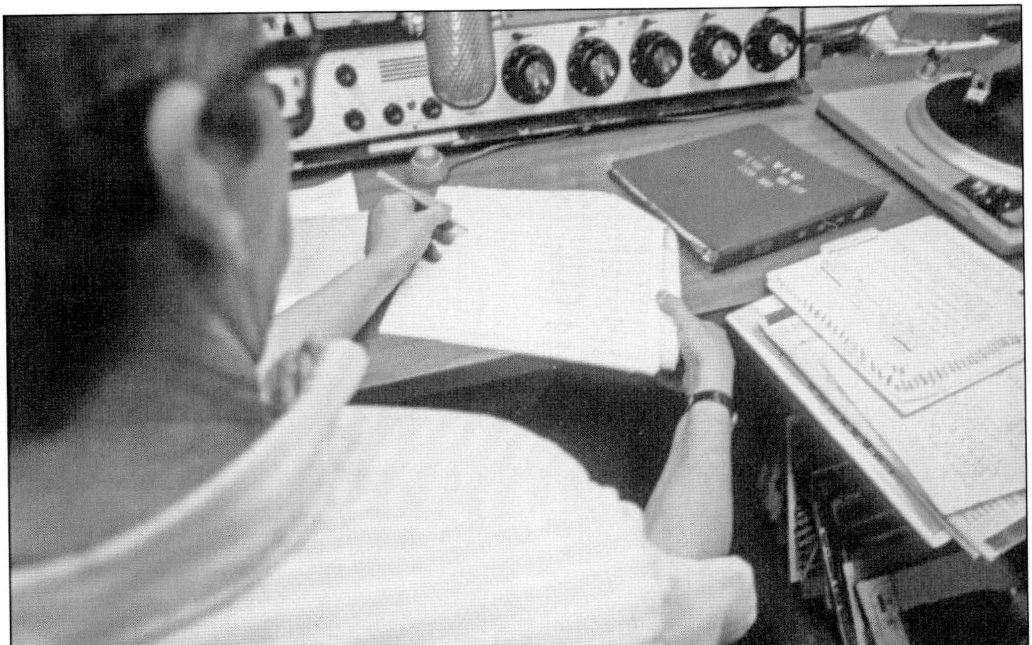

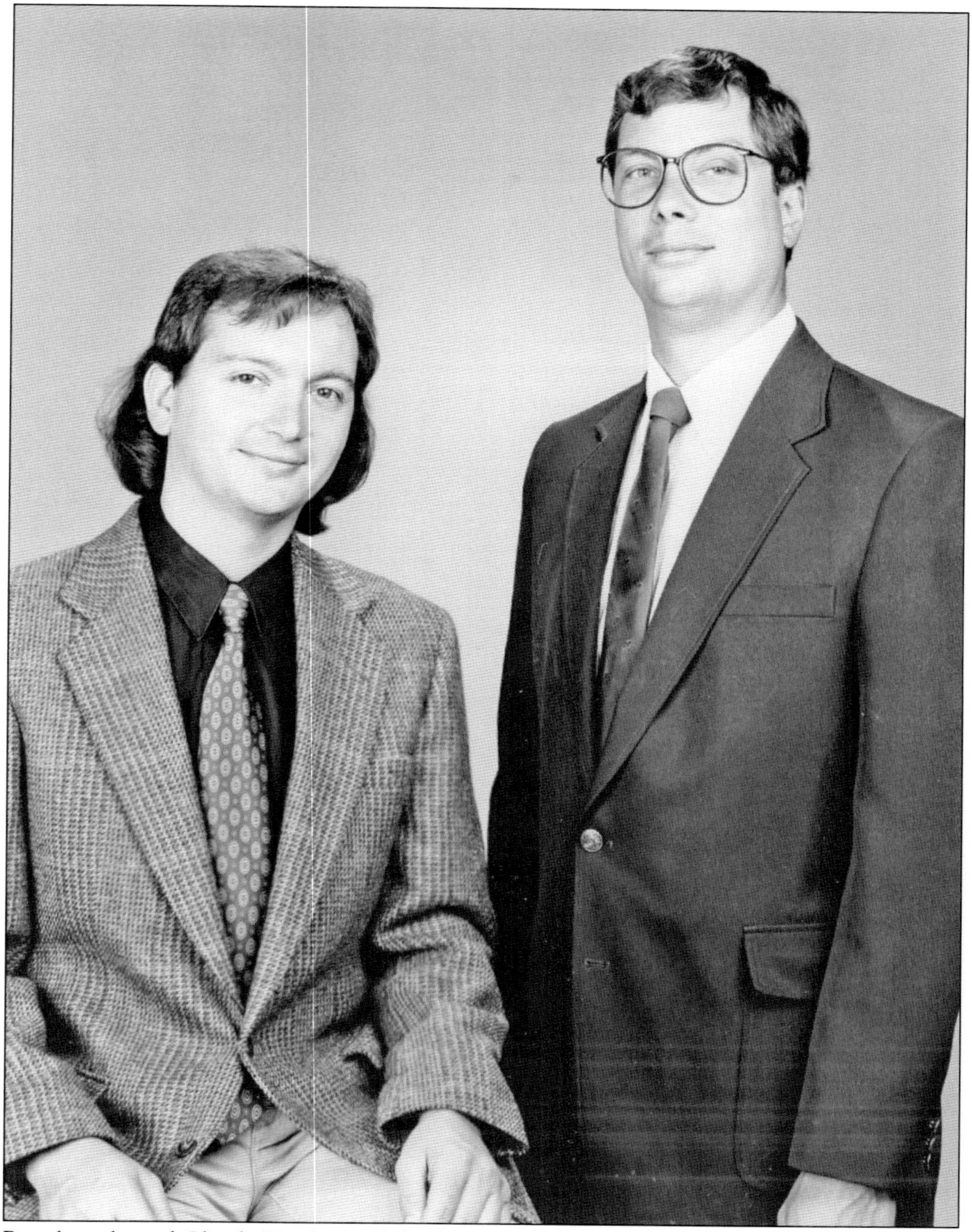

Broadcast legend Claude Tomlinson left pretty big shoes to fill. Wilhite and Wall were hired to do just that and took the morning show in a different direction, keeping the ratings high during their tenure. From left to right are Darren Wilhite and Tim Wall. (Photograph courtesy Wayne Bledsoe.)

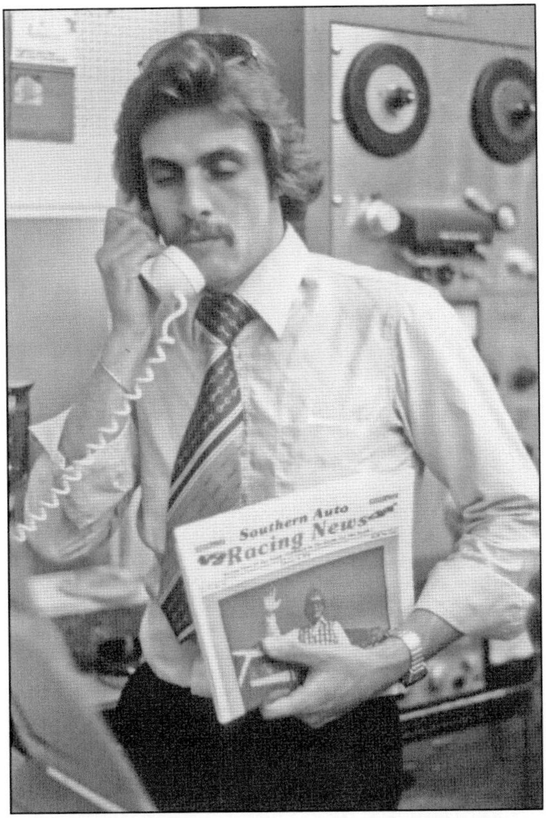

Jimmy Vineyard was a popular weekend announcer for the station. His Sunday night show that featured the latest bluegrass bands was a hit in the early 1980s as bluegrass began to find an independent voice of its own.

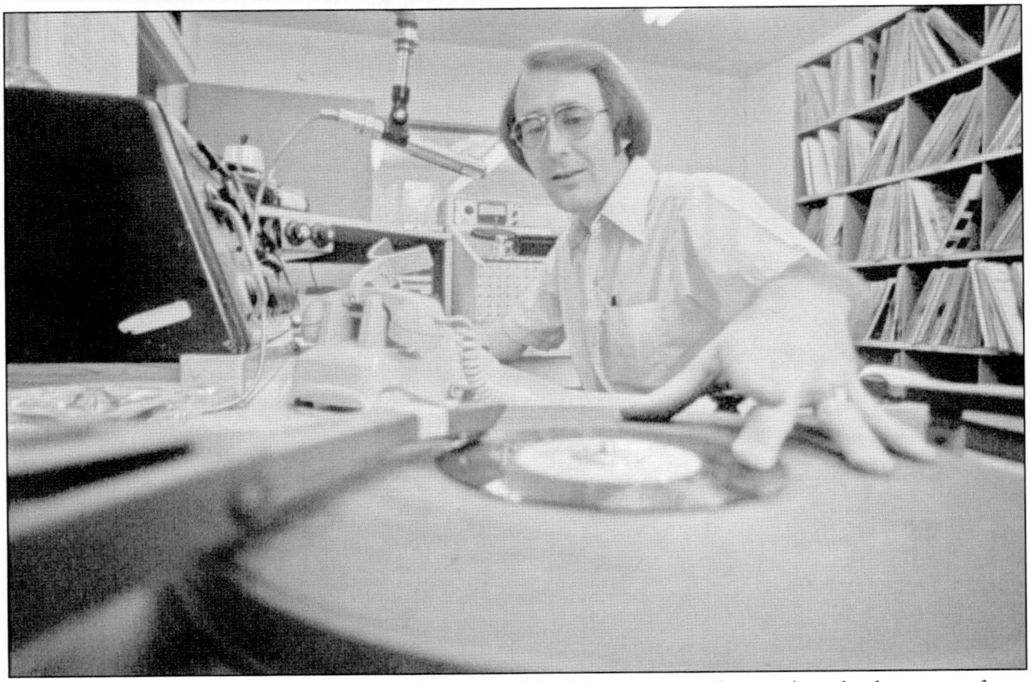

Bob Catron puts another record on the turntable. The station changed with the times from albums to carts to digital.

Three

THE ORIGINAL INFORMATION STATION

From the 1960s to the early 1980s, radio news was at its peak. Cable news networks were still in the imaginations of their founders, and the Internet was in its early stages of development. Local television stations rarely if ever in those days broke into soap operas or other programming for breaking news. Newspapers had long abandoned printing "extra" editions. If people wanted to keep up with breaking news or other developments before the 6:00 p.m. television newscast, they had to rely on radio. Recognizing the importance of local news and the need for accurate, immediate reporting in the area, Jim Dick decided to dedicate the same passion that had worked so well with the announcing staff to establishing a professional news operation.

Norman Majors was working with Kenny Dearstone on the morning drive at the time, and Jim Dick tapped him to do it. Majors, who is now senior vice president of the University of Tennessee Hospital in Knoxville, became WIVK's first news director. He and Jim Dick went to the basement and built what would become one of the most relied-upon information sources in the city.

In years past, disc jockeys and news readers would merely comment informally on news of the day, but adding a dedicated news department plus a couple of part-timers and stringers, the station found another way to not only serve the public, but also to generate ratings and revenue.

"We put in a few tables, Jim brought in some phone lines and a wire service, and that's pretty much how we started," said Majors. "We kept up with WNOX, who was the leader, and had a pretty good sound, which gave us something to work with. Like anyone who's worked there, I still have a fondness for the station."

With WIVK easing into the news business, new ideas were born that would change the sound of Knoxville radio forever. Giving listeners what they wanted to hear was a job WIVK had done professionally for 20 years, and Jim Dick wanted to make sure listeners also got when they needed to hear from state and local news. In 1973, Rogersville native Mike Hammond was hired into the news director's slot, and he reorganized it into a first-rate modern operation and pushed the station's news department to new heights, adding the first "Eye in the Sky" traffic reports in the city and investigative and feature reporting.

Reporter and anchor Jean Ash, who joined the staff in 1975, specialized also in lengthy series, which were unique to WIVK in the market. "Mike would give me a week off from regular reporting and anchoring duties to develop feature series on a wide variety of topics, Ash said. "Some of these included 'Fear of AIDS' and in-depth looks at the Cherokee Indian Reservation on the other side of the Smokies."

At its peak, WIVK's news department was larger than most in major markets. It boasted five full-time reporters and anchors plus a full-time sports director and a part-time sports reporter. WIVK News ran like a major market operation, even sending reporters to the national presidential conventions in 1988.

Weather is also a mainstay of the station. In winter months, school kids in East Tennessee tune in for the "SnowBird Report" to hear if the snowfall the night before had given them a day off from school. Public affairs, which had always been required by FCC regulations, also took on a new sound. Call-in shows allowed the listener to interact with the announcers and informed East Tennesseans on local and state affairs. *Sound-Off* remains one of the longest-running public affairs programs in radio.

WIVK's news department became a vital part of daily life in East Tennessee, and sports, which had always been a part of the radio station, now started developing a news angle all its own. Broadcasts of the University of Tennessee football games were tuned in to on radios all over the region, and the station put its stamp on UT sports, even being one of the earliest supporters of the Lady Vols basketball team before they started climbing to one national championship after another.

In short, WIVK did for news what it did for country music in East Tennessee and netted the most prestigious national awards in the broadcast business. Whenever a story was breaking and television crews and newspaper photographers shot pictures in interviews, the WIVK microphone flag was always on the podium. In 1990, the Associated Press recognized WIVK as the Best News Department in Tennessee, and three of its members would be Associated Press Broadcasters of the Year.

WIVK news coverage goes far beyond the listening area it is assigned. As a part of the American Broadcasting Network, when tragedies and big stories occur that affect listeners, the station has sent reporters to cover them—from hurricanes, earthquakes, and even war fronts when local soldiers were called to serve. In 2003, the author, then a reporter, and on-air personality Ted "Gunner" Ousley became the first WIVK team to travel to a combat zone when they were shipped in with the local U.S. Army Reserve's 489th Civil Affairs Battalion to Iraq.

It is all part of the news philosophy of WIVK founder Jim Dick that what concerns the listener concerns the station. The Independent Voice of Knoxville added a new standard of excellence to its daily broadcasts and a new group of independent voices that would be welcomed into homes across East Tennessee.

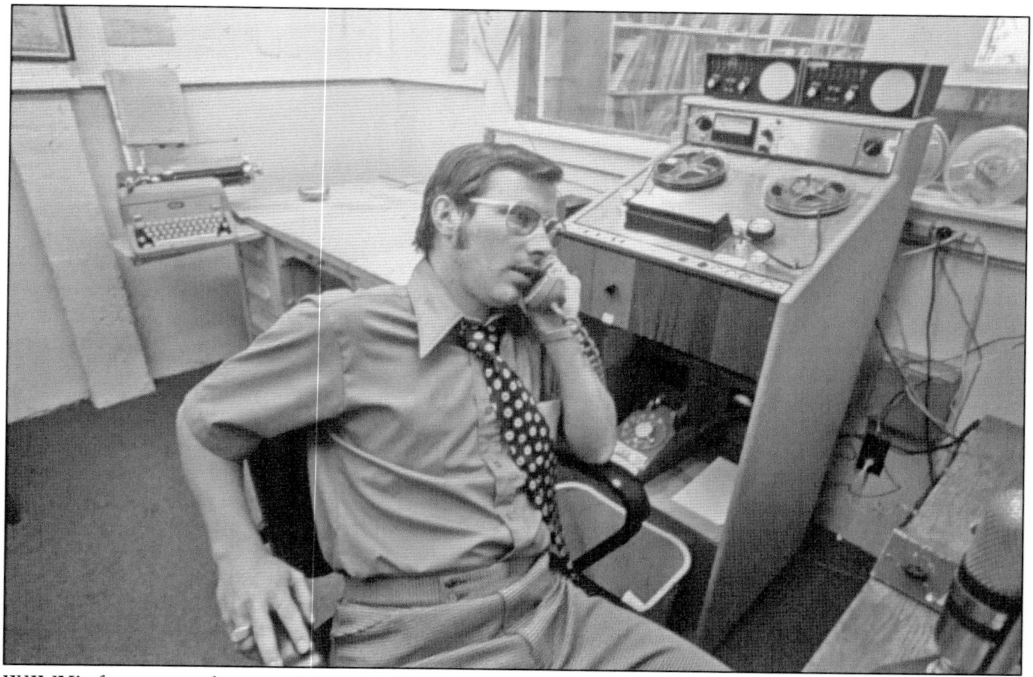

WIVK's first news director, Norman Majors, is pictured on the phone. The uniqueness of radio made the telephone a viable option in getting interviews from news makers of the day and is still a staple of news broadcasts.

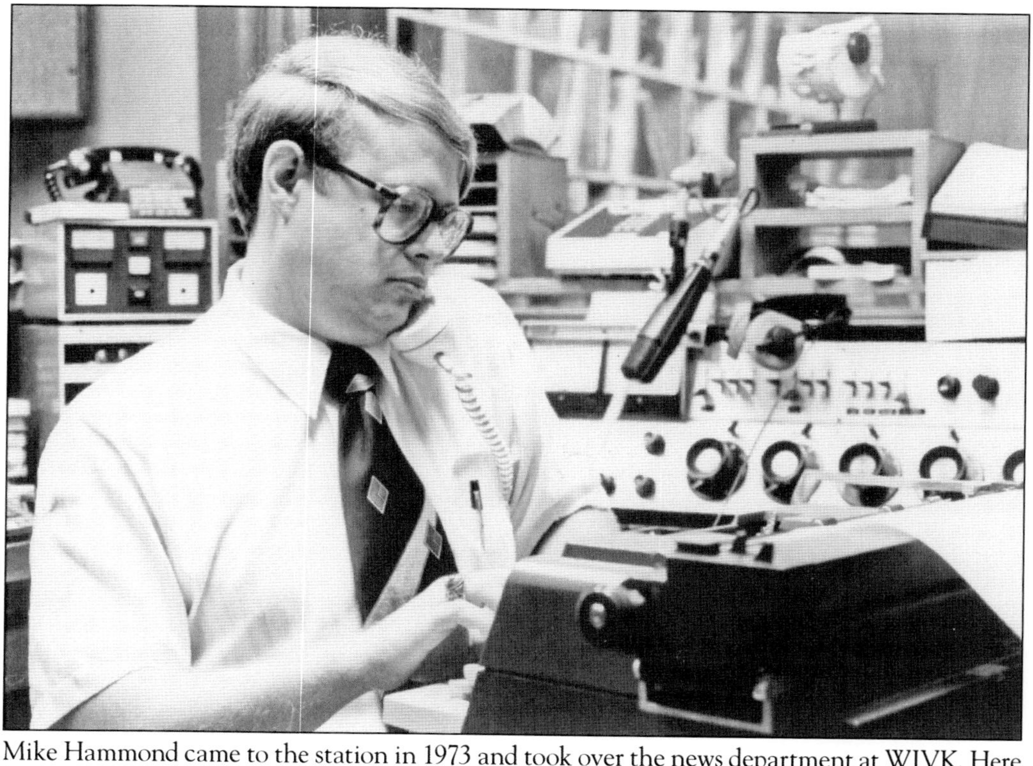

Mike Hammond came to the station in 1973 and took over the news department at WIVK. Here he is pictured in the newsroom studio preparing a news broadcast at the typewriter.

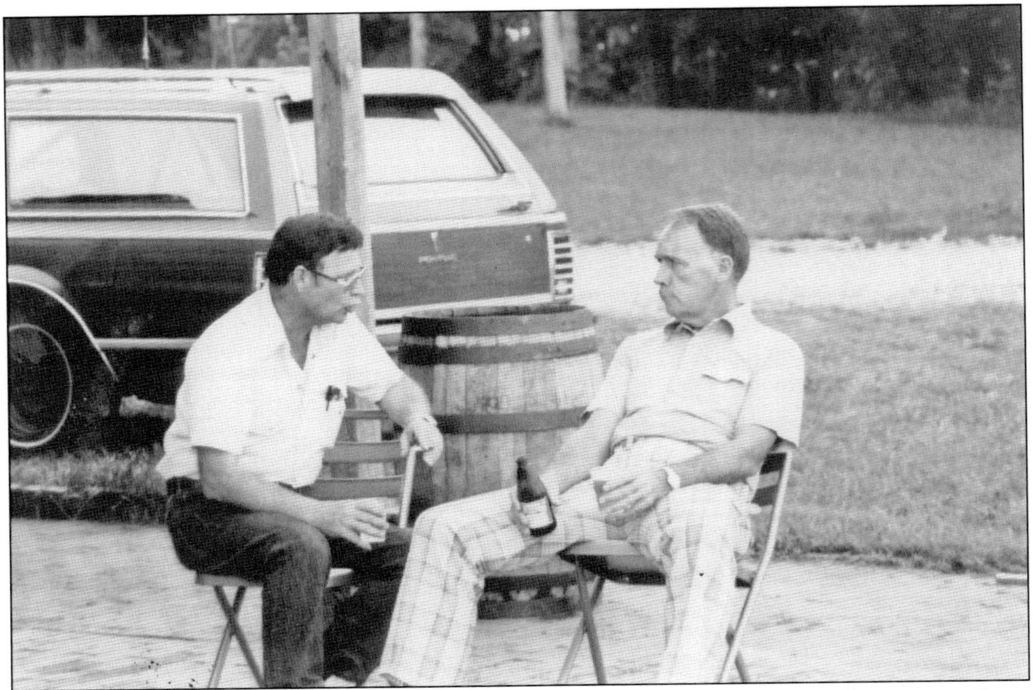

A popular feature of WIVK's Information Station format was its aerial traffic reports. From left to right are station owner Jim Dick and Knoxville's first traffic reporter, Knoxville police sergeant Wayne Bell. Bell would become the city's first voice that commuters listened to every morning and afternoon on their commute. Below is an aerial view of Knoxville from a traffic reporter's perspective.

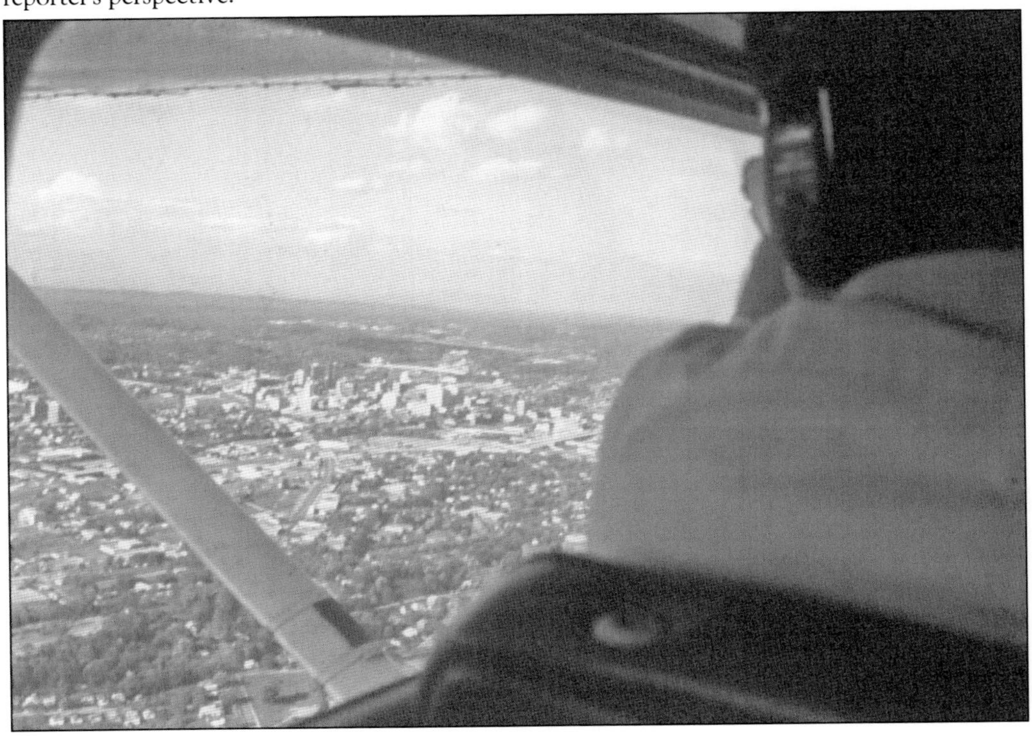

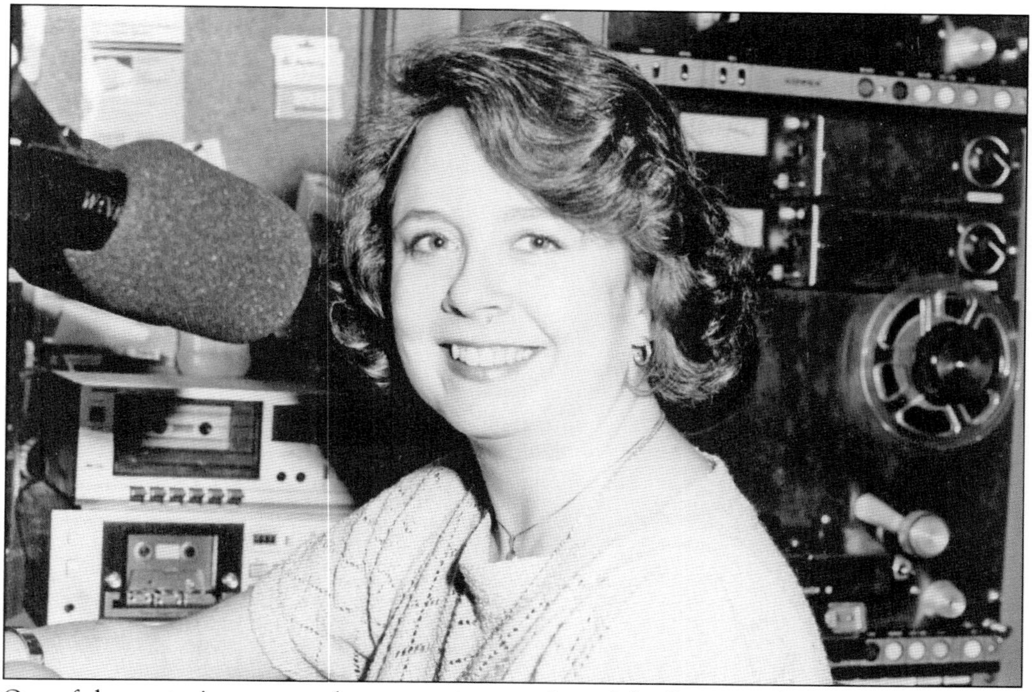

One of the station's most popular news voices was Jean Ash. She started in 1975 at the station and is regarded as one of the most professional news people in the business, earning the coveted AP Broadcaster of the Year in 1987 for her work. She would go on to become a vital part of East Tennessee journalism through her efforts in the Society of Professional Journalists (SPJ). The SPJ was viewed as a newspaper organization, and she worked hard to successfully make it more responsive to the needs of broadcasters.

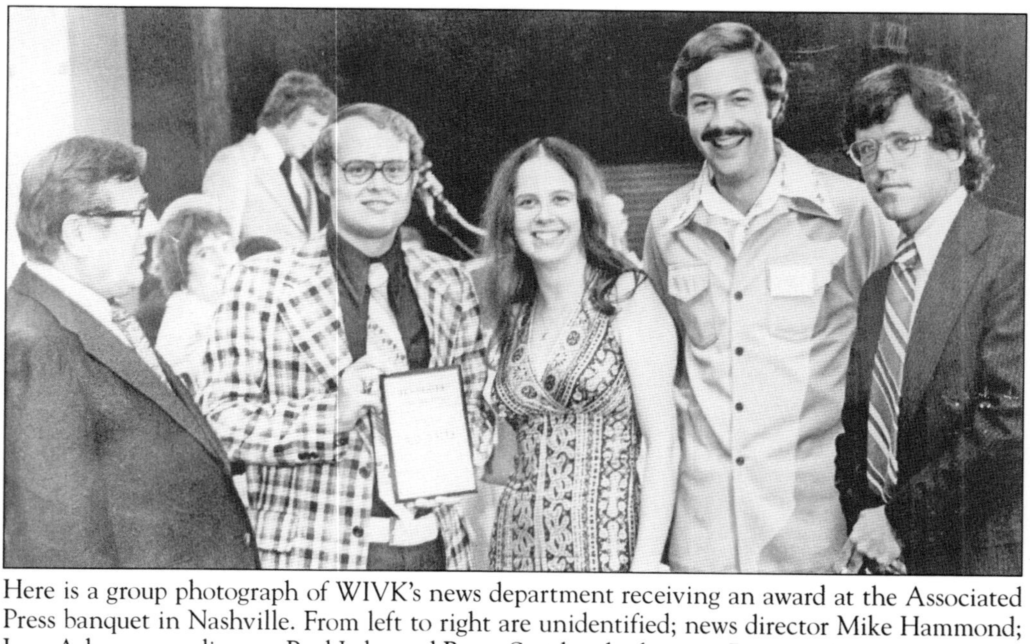

Here is a group photograph of WIVK's news department receiving an award at the Associated Press banquet in Nashville. From left to right are unidentified; news director Mike Hammond; Jean Ashe; sports director Paul Lyle; and Barry Smith, who became Barry Nolan and worked as a reporter for *Hard Copy* and *Extra.* (Photograph courtesy of Jean Ash.)

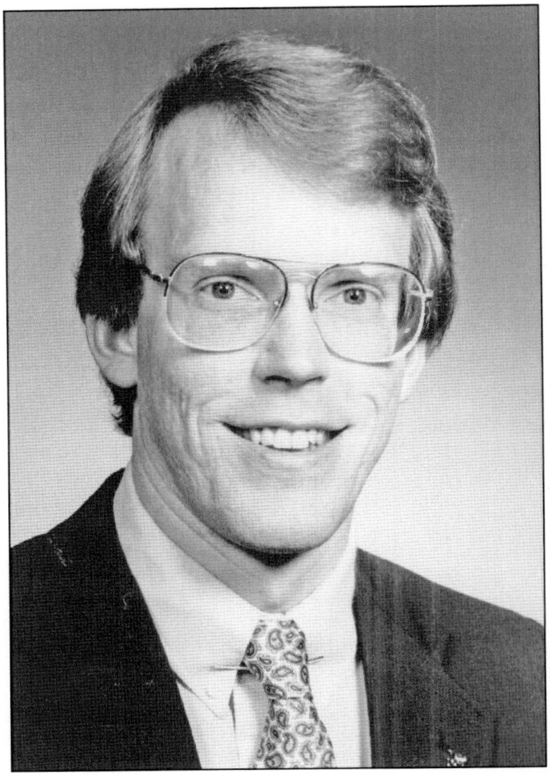

Beecher Bailey was a long-time reporter for the station who covered breaking events and news. Below is news reporter Tonya Stout Brown, who also hosted an afternoon spoof of the news called *Afternoon Sleaze*.

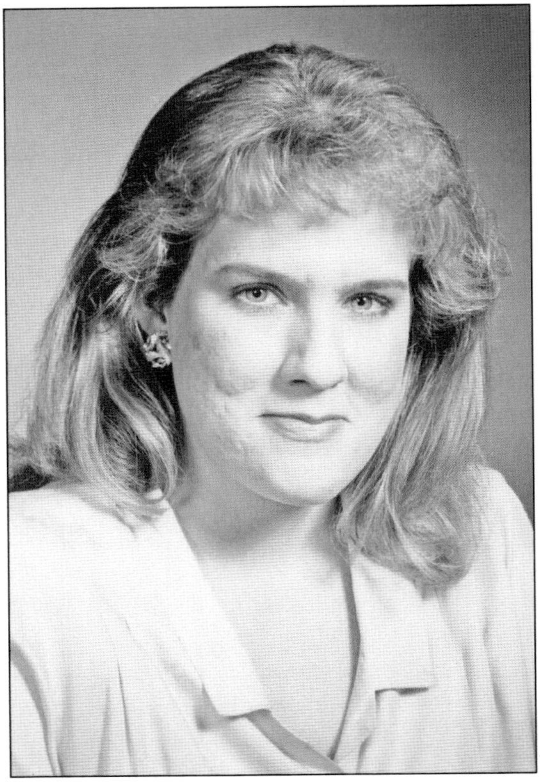

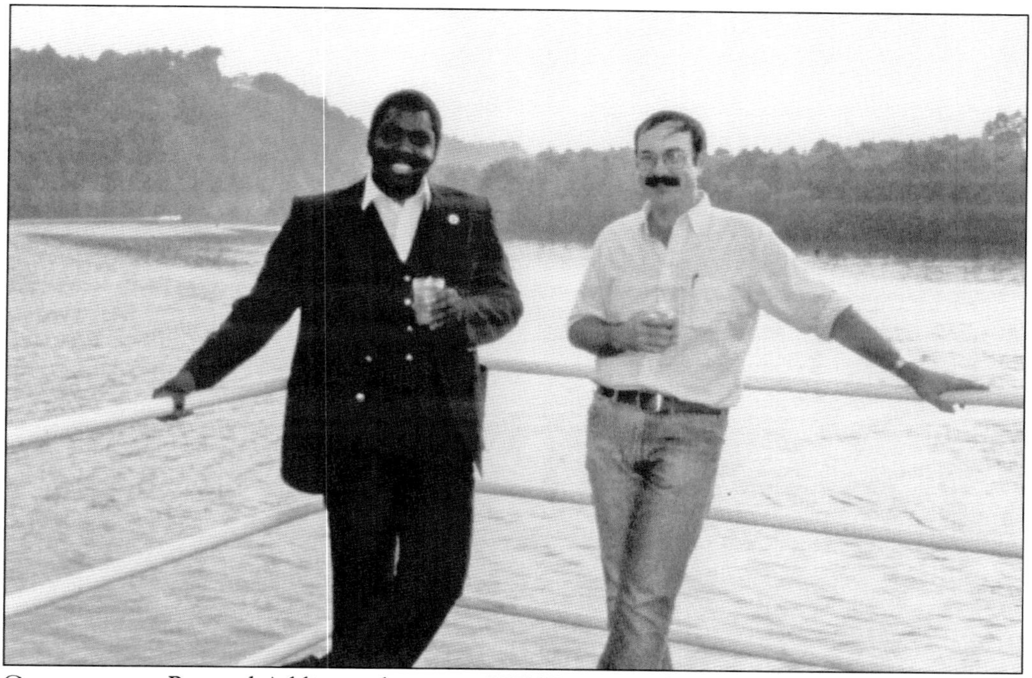

Court reporter Bernard Addison relaxes at a WIVK party with Steve Ash (right), the husband of news anchor Jean Ash, on a riverboat. (Photograph courtesy Jean Ash.)

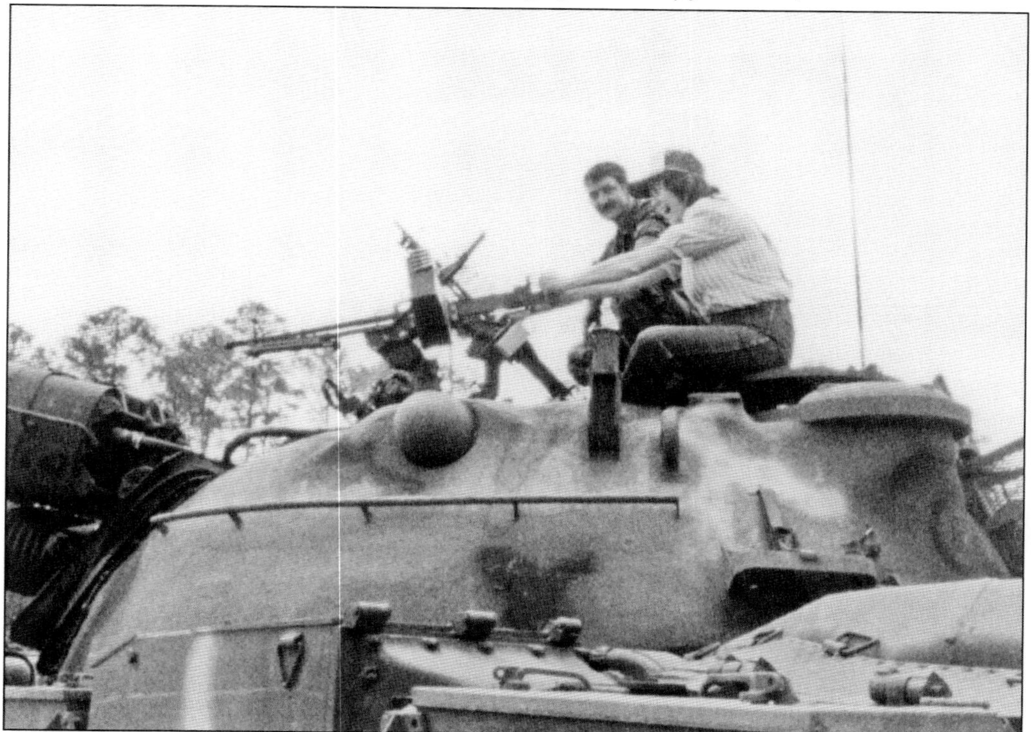

The military service of East Tennesseans never went unrecognized by WIVK. The photograph above shows Jean Ash on a visit to Fort Stewart, Georgia, to report on the National Guard training operations. Here she sits on a tank at the training ground with one of the guardsmen.

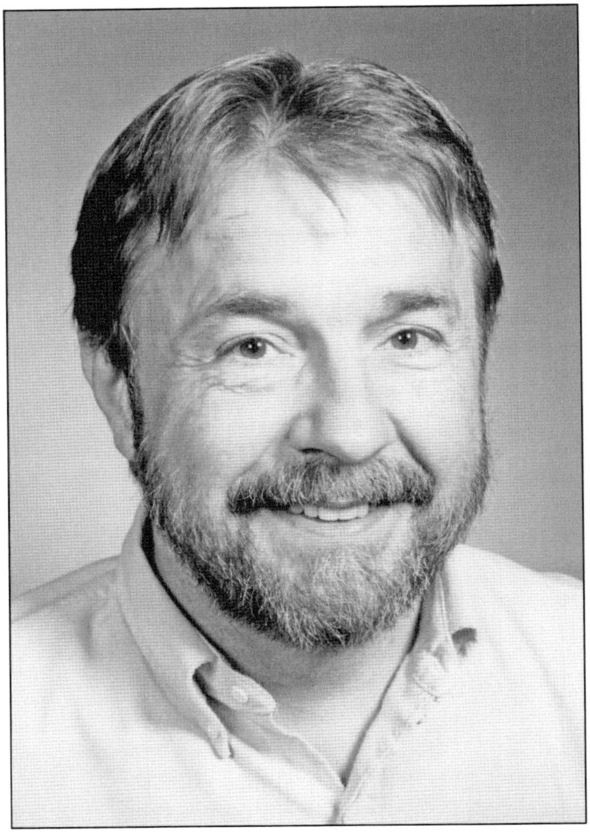

Following Wayne Bell's departure from WIVK, Dan Bell went to work for WIVK as a traffic reporter. He remained with the station until he went to work for the Knoxville Transit Company.

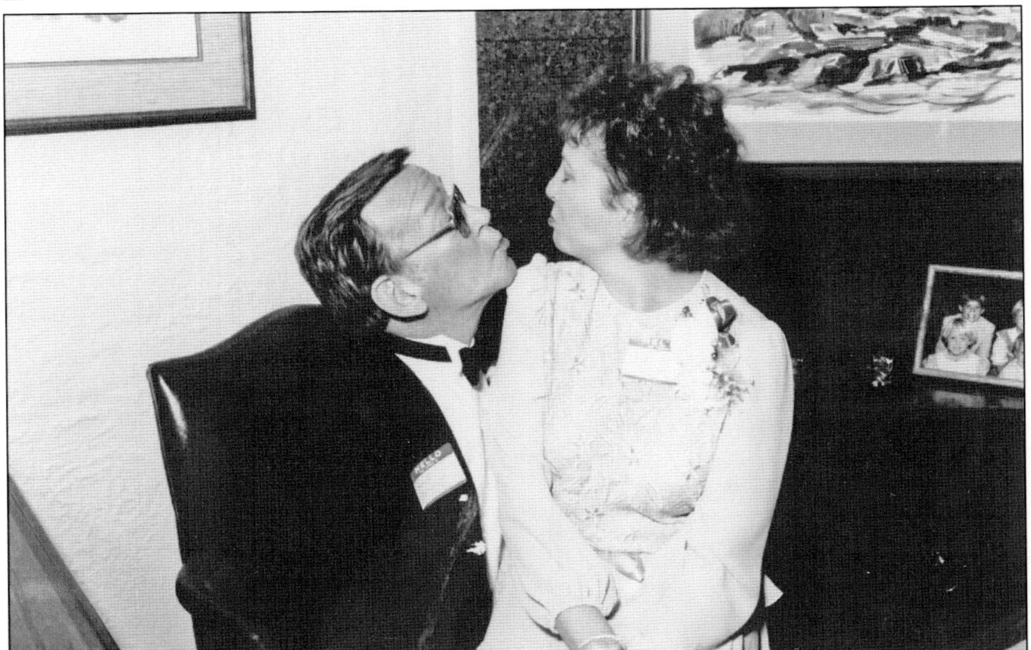

WIVK news anchor Jean Ashe and Claude Tomlinson fake a kiss for the camera at the 30th anniversary party. (Photograph courtesy WIVK.)

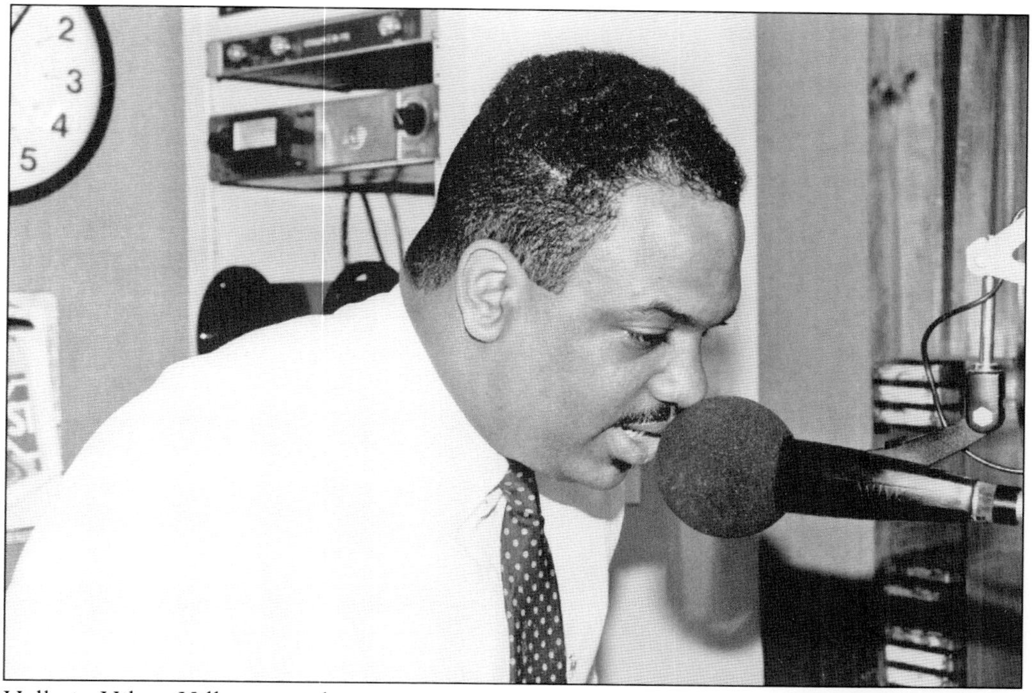

Hallerin Hilton Hill was another popular news reporter for WIVK and covered a number of big stories during his time at the station. He is still in the building, as he went on to host the popular *Triple H Morning Show* on WNOX, which is now a sister station of WIVK.

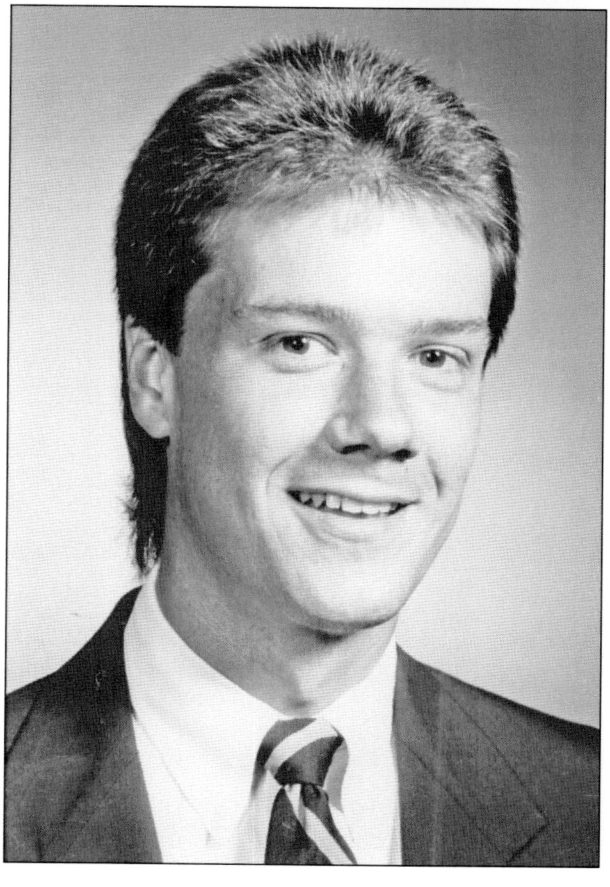

News anchor Channing Smith started at the station just before the 1982 World's Fair. He is still at WIVK, anchoring the newscasts for the *Andy and Allison Show*. He was a longtime news reporter for the station before moving into the morning anchor slot.

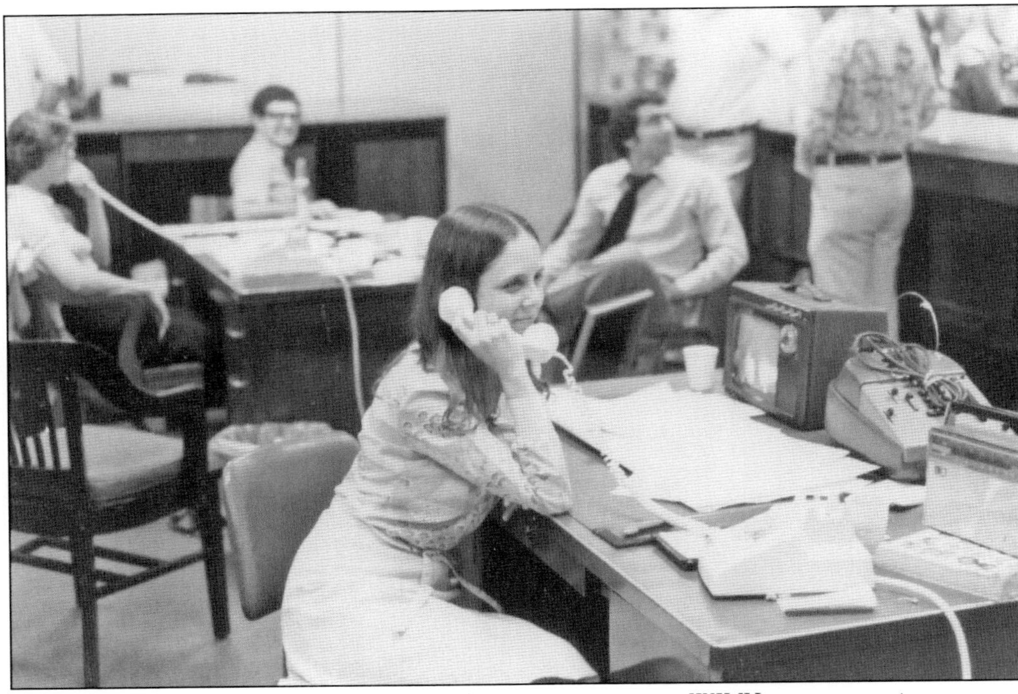

Jean Ash works election coverage at the election commission. WIVK news operations were so relied upon by other news organization that the *Knoxville Journal* and local television stations used WIVK to get precinct voting numbers.

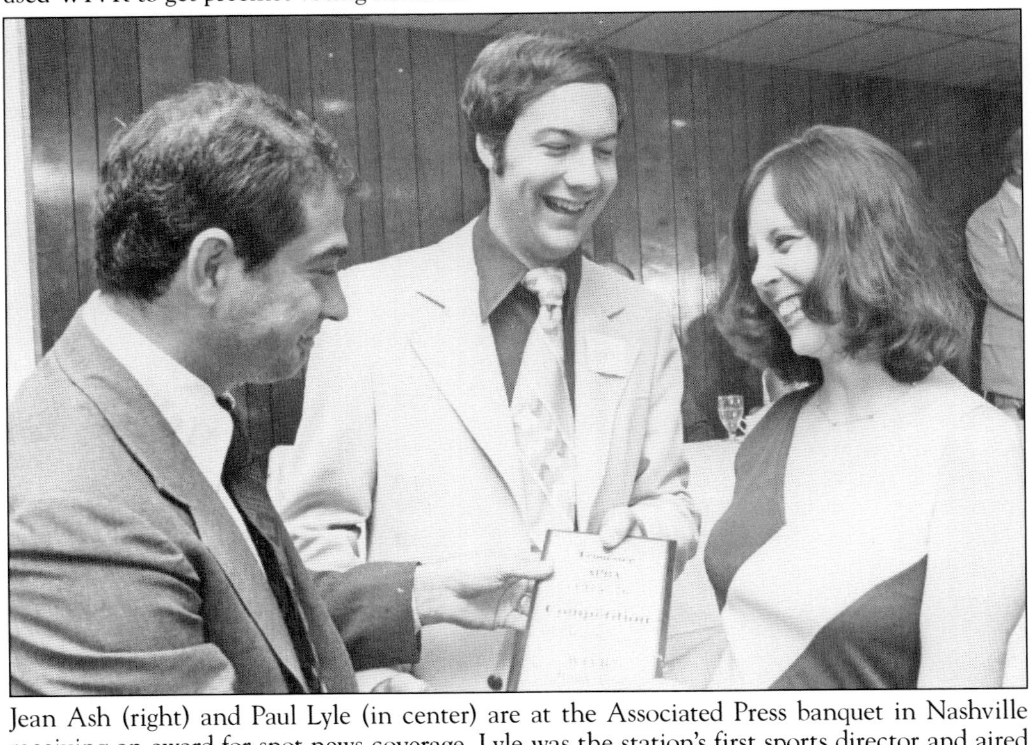

Jean Ash (right) and Paul Lyle (in center) are at the Associated Press banquet in Nashville receiving an award for spot news coverage. Lyle was the station's first sports director and aired the first-ever sports talk show in Knoxville. (Photograph courtesy Jean Ash.)

Announcer Bob Catron stands at the Associated Press wire machine at the station. The machine stayed in operation at WIVK until early 1992, when the wire service was integrated into a new computer system.

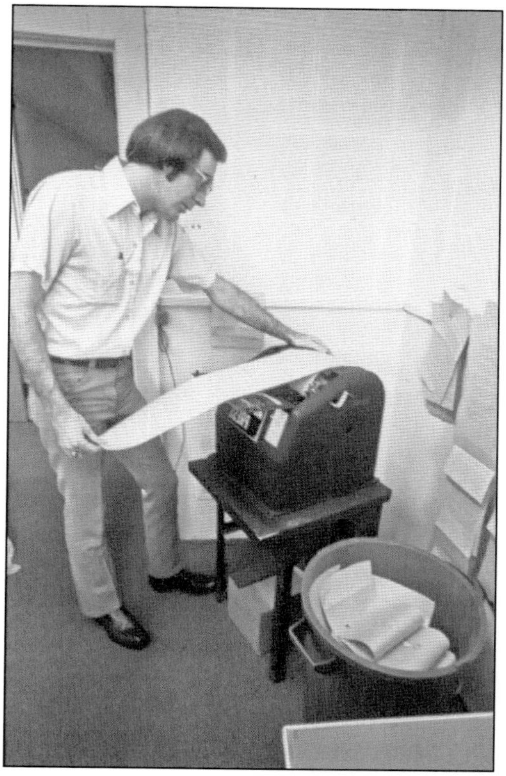

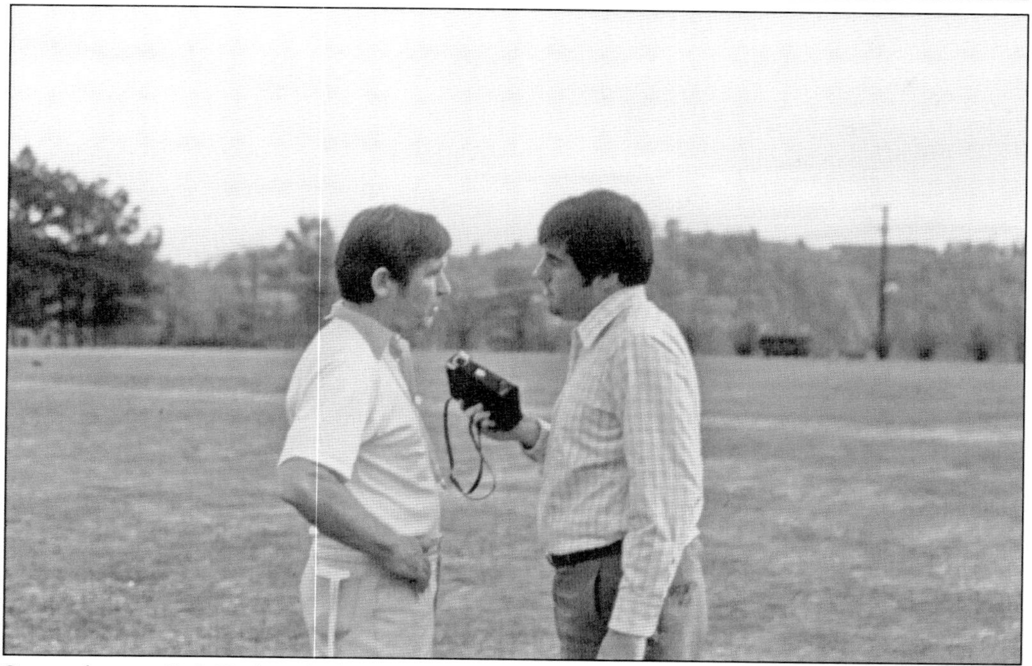

Sports director Bob Kesling (right) interviews UT football coach Johnny Majors. Kesling started at the station while attending UT and took over after Paul Lyle left. He later became the sports anchor at WBIR TV and is the "Voice of the Vols" today, broadcasting the UT football action to a network of radio stations throughout the Southeastern Conference area.

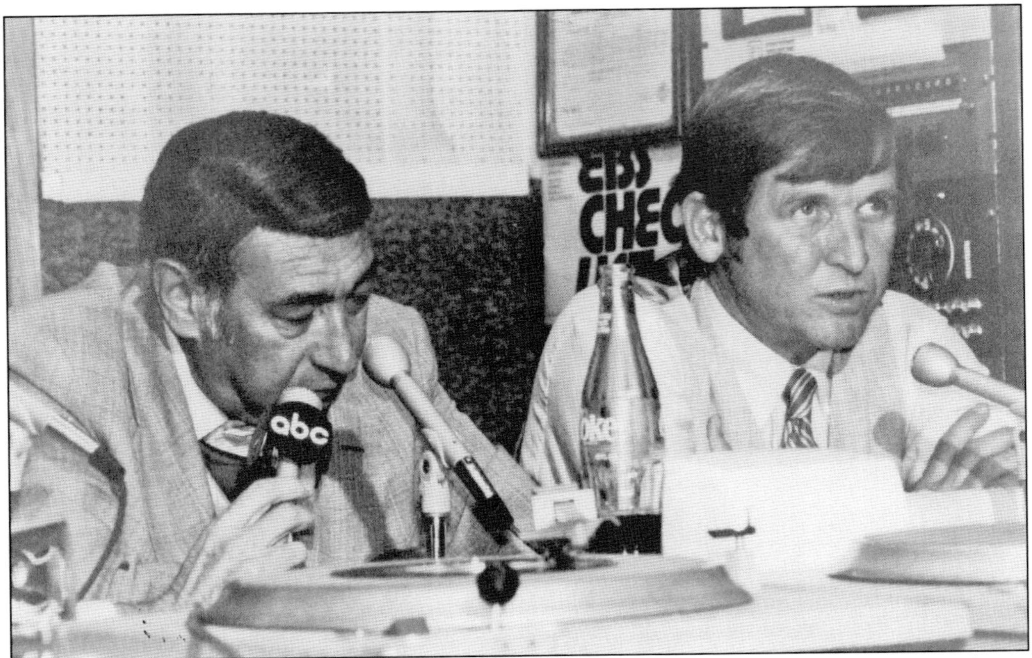

From left to right are broadcast sports legend Howard Cosell and UT football coach Johnny Majors following a UT football game. WIVK's wall-to-wall coverage of UT football is one of its biggest ratings winners. (Photograph courtesy Foster Arnett.)

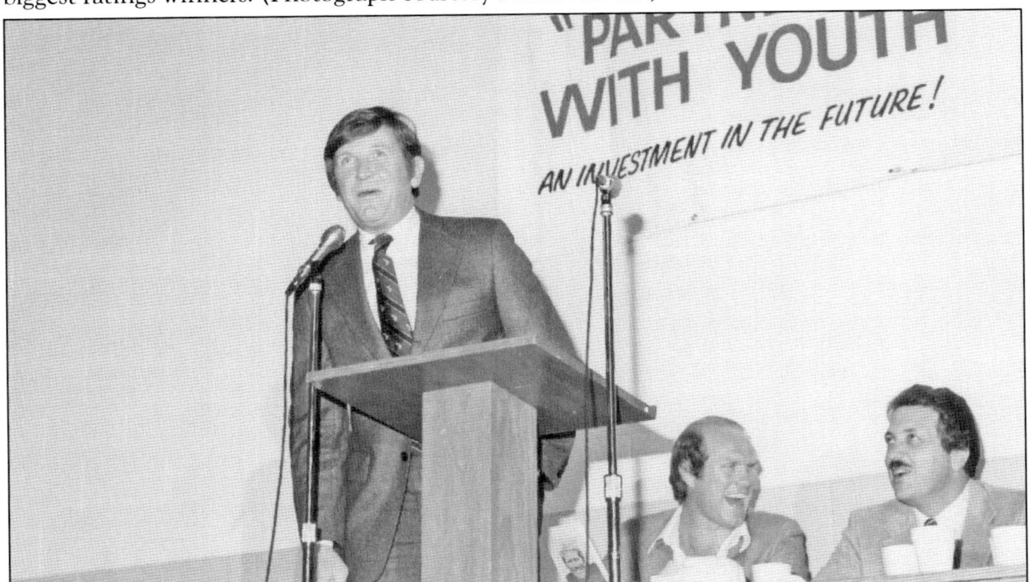

WIVK sports aided a number of causes in the city with its ability to bring in sports stars for charitable causes. Here UT football coach Johnny Majors speaks at an event. From left to right in back are WIVK general manager Bobby Denton and former Pittsburgh Steelers quarterback Terry Bradshaw. (Photograph courtesy McClung Historical Collection, Knox County Public Library System, Knoxville, Tennessee.)

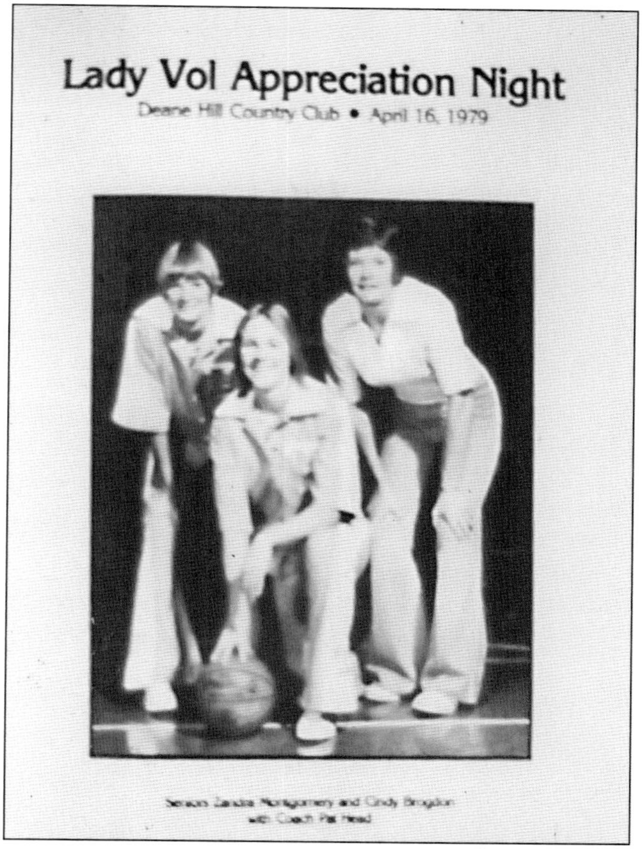

WIVK was one of the first supporters of the UT Lady Vols basketball team. Here is a copy of the program on Lady Vols Appreciation night. In the photograph below, Lady Vols coach Pat Head Summitt is seen speaking at the event.

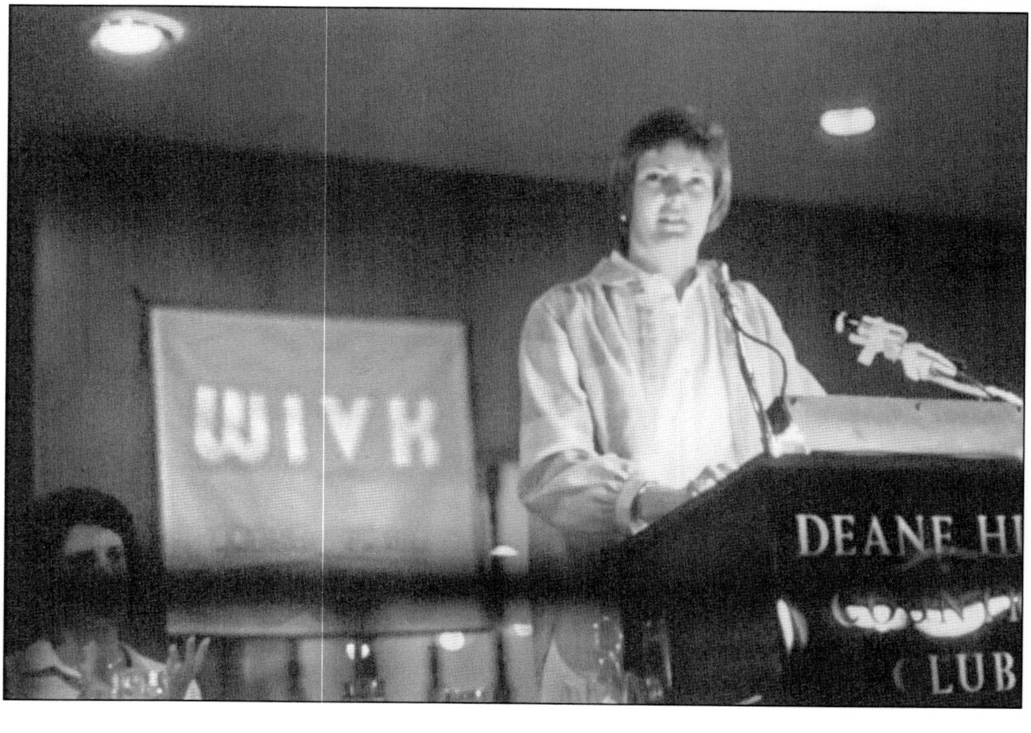

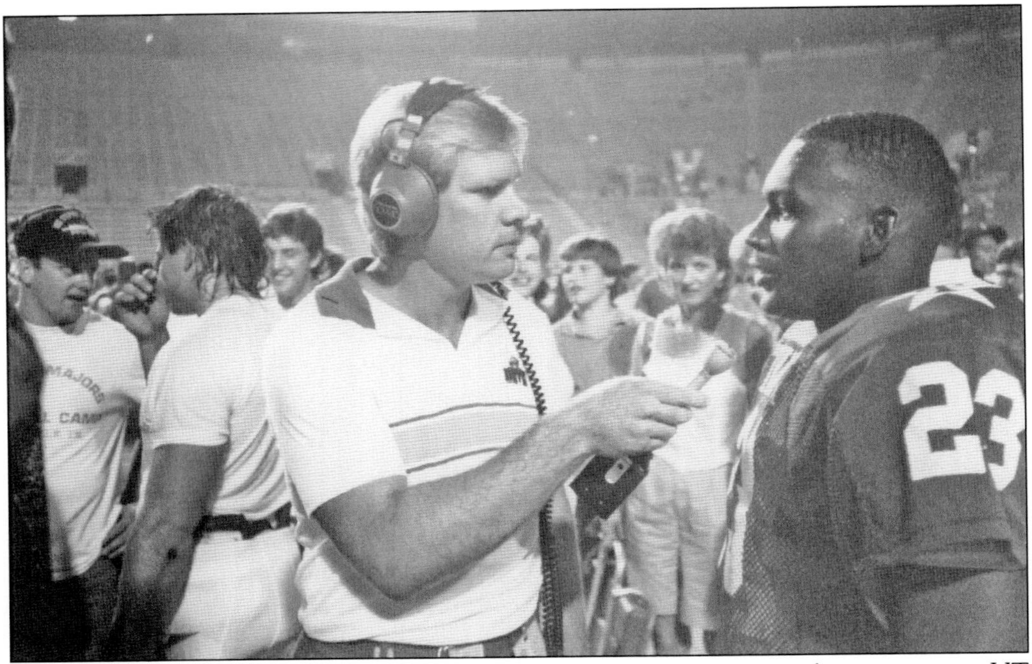

Alan Williams was another WIVK sports director at the station. Here he is seen at a UT football spring game interviewing one of the players. Williams went on to become a television sports anchor in the city and is today the news anchor at WVLT TV. (Photograph courtesy Foster Arnett.)

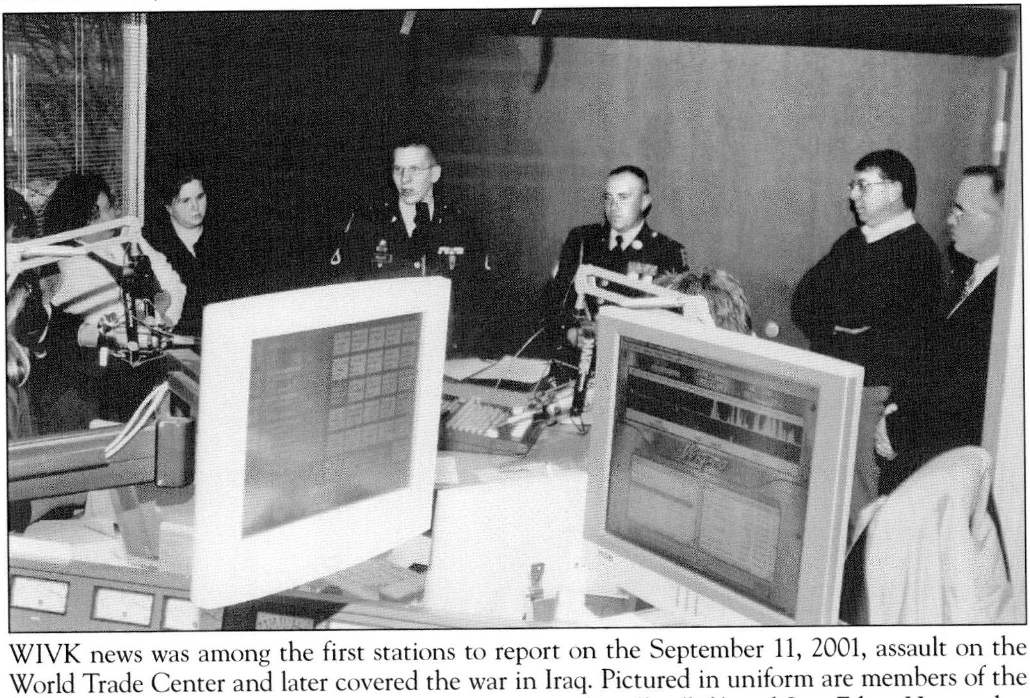

WIVK news was among the first stations to report on the September 11, 2001, assault on the World Trade Center and later covered the war in Iraq. Pictured in uniform are members of the U.S. Army's 507th Maintenance Company, Pfc. Patrick Miller (left) and Spc. Edgar Hernandez, who were among the first to be captured by Iraqi militants. The former prisoners of war served as grand marshals of the WIVK Christmas parade. (Photograph courtesy Ed Hooper.)

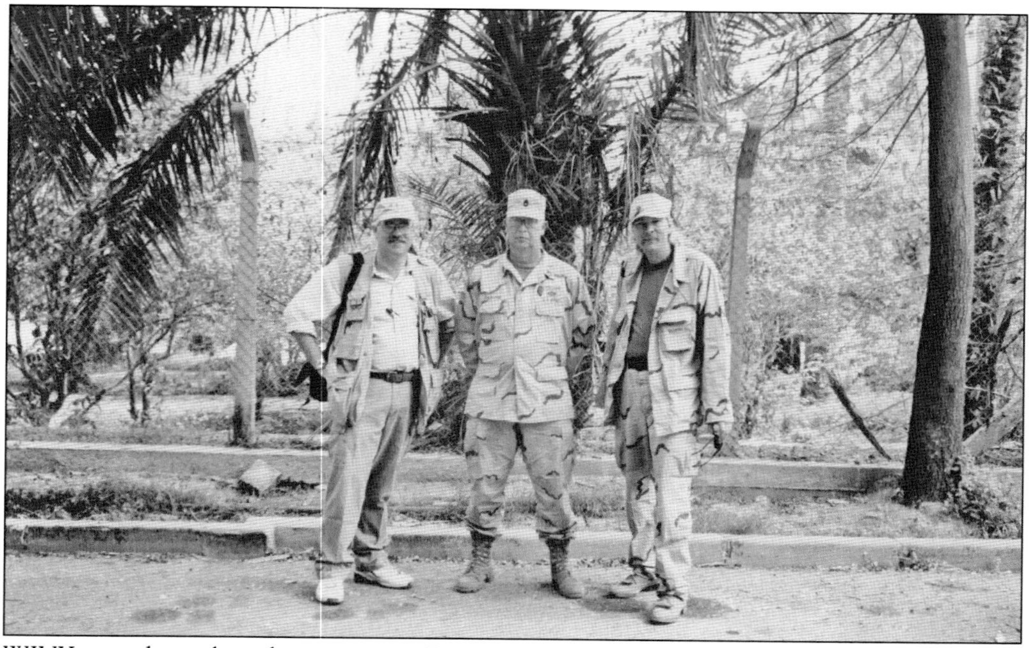

WIVK was the only radio station in East Tennessee to embed a news team with the U.S. Army Reserves 489th Civil Affairs Battalion (CABN) in Iraq. From left to right are afternoon announcer Ted "Gunner" Ousley, Sgt. Joe Raeburn, and author—then reporter—Ed Hooper. News reporters were gathered in the field and later produced on a laptop computer. Live reports were filed with a satellite telephone from the field. (Photograph courtesy Ed Hooper.)

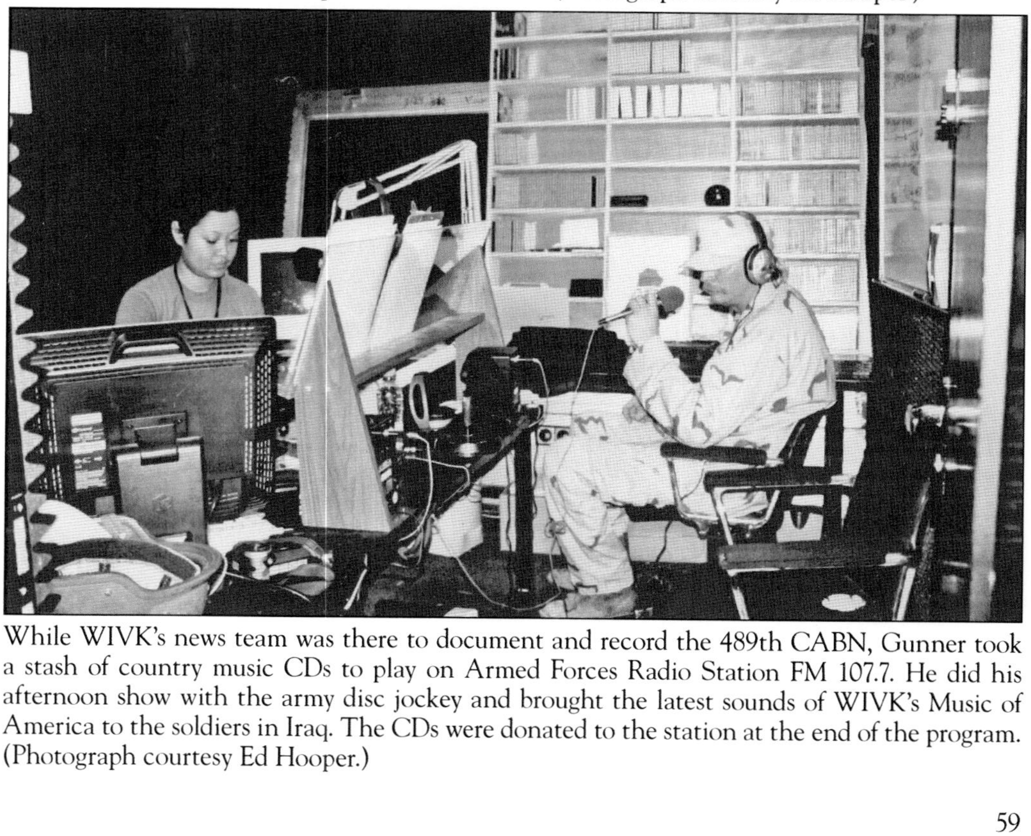

While WIVK's news team was there to document and record the 489th CABN, Gunner took a stash of country music CDs to play on Armed Forces Radio Station FM 107.7. He did his afternoon show with the army disc jockey and brought the latest sounds of WIVK's Music of America to the soldiers in Iraq. The CDs were donated to the station at the end of the program. (Photograph courtesy Ed Hooper.)

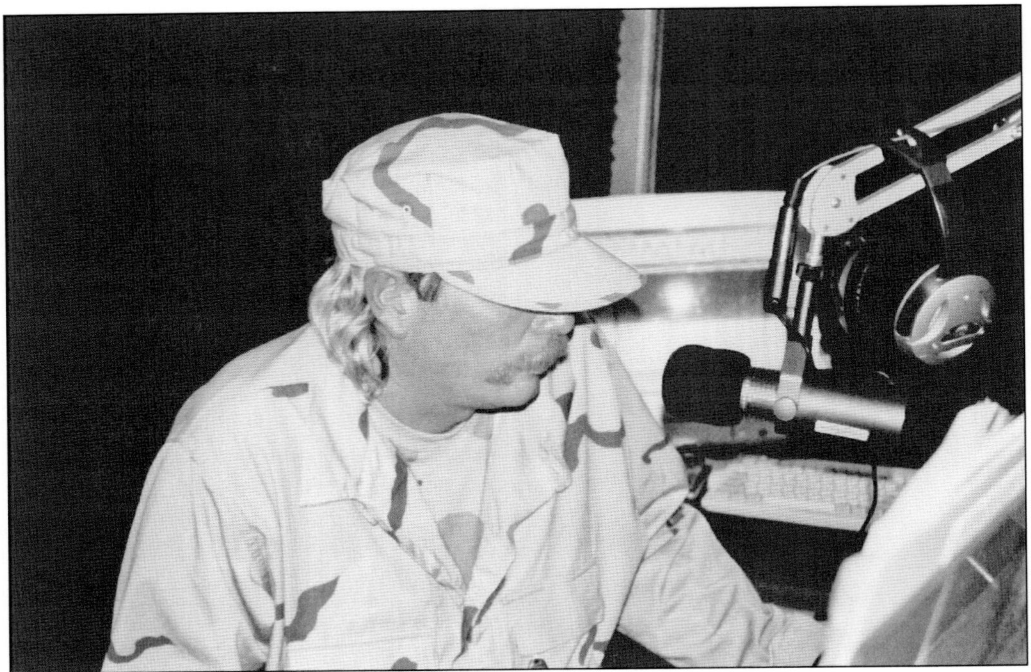

Gunner delivered messages from country music entertainers across the United States to the troops in Iraq during a two-hour show that turned the Armed Forces Radio station from its "Bright Adult Contemporary" format to the Music of America. (Photograph courtesy Ed Hooper.)

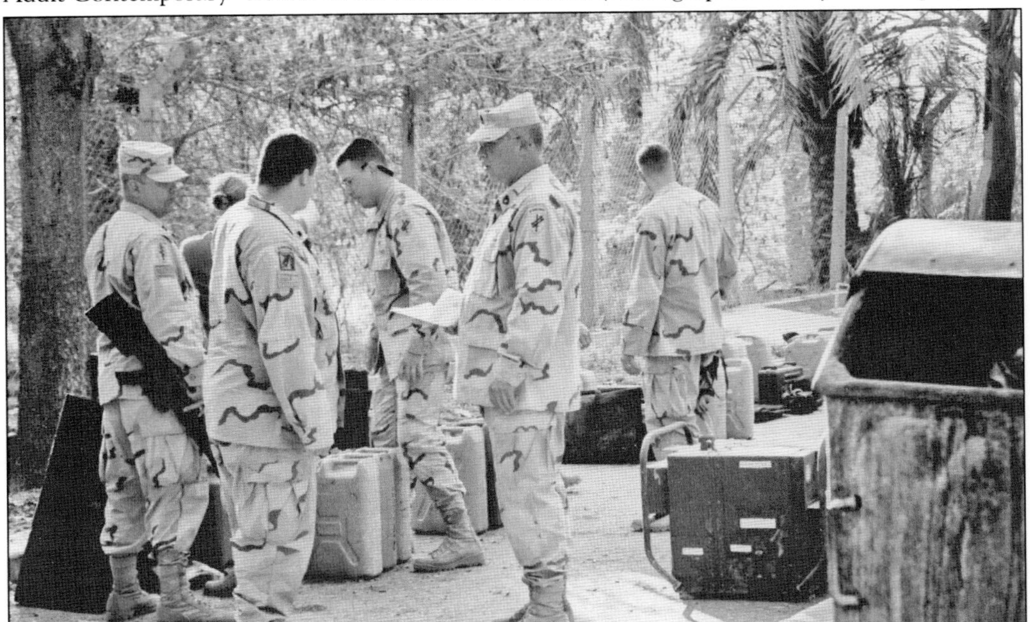

Here Sgt. Joe Raeburn (second from right) of the 489th CABN organizes the move to the forward operating base where the unit was headquartered. WIVK's Gunner and the author stayed with the unit from the minute they left the ground until they received their final orders to make their home. The civil affairs battalions were key in helping reestablish Iraqi control of their government and resources. (Photograph courtesy Ed Hooper.)

Four

FORGING A HERITAGE FORMAT

Since Jim Dick first founded WIVK, the station, with the exception of a three-year period, had remained trued to its original musical format that attracted listeners from across East Tennessee. As times changed and the station evolved from its small Gay Street studio to a powerhouse, the once-acoustic sounds of "hillbilly music" had come a long way as well. It was now regarded, as the station motto proclaimed, as "the Music of America."

The founder never expected the station to be regarded as a heritage station one day. WIVK was more than a jukebox for East Tennessee. It was a living, breathing entity that helped build country music from the ground up. In fact, the station holds seven Country Music Association Station of the Year awards, more than any other in the nation.

It would be next to impossible to name a country music entertainer, especially in the early years, that did not go through WIVK at one time or another in their career. There are many whose careers actually began on its airwaves. The unsung heroes of a radio station are the program directors who select the music that gets played on air and the performers that make the cut. It can be the most dangerous job because that position can determine if a station fails or succeeds.

It was not only what came across the airwaves of WIVK that made country music entertainers stars, it was the performers the station brought to town. When the station changed from its Top 40 format to go back to its country music roots, Jim Dick decided it would be good to get back to the live performance side of the business. It was a great jump for the disc jockeys, who would get to announce the acts at Knoxville's Civil Coliseum. The WIVK *Shindig* became one of the most recognized entertainment forums in East Tennessee. WIVK hosted the concerts six or seven times a year, and the station would devote its time and staff to producing what became an East Tennessee pastime. One of the first ones featured a new singer who had left the popular American folk band the New Christy Minstrels to make a go of it on his own. His name was Kenny Rogers. The other names read like a who's who of American country music—Conway Twitty, Loretta Lynn, George Jones, and others paraded across the stage.

The days of the *Shindig* were some of the most memorable for the station, and it never completely left them behind. WIVK is still known for its promotion of rising stars and its support of new talent. It mixes the old with the new and blends them into the sounds that have now been the music of choice for generations of East Tennesseans.

This is the reason WIVK became part of the country music family and why its format has remain unchanged for more than 40 years. This is the backbone of the station's success, and many of its employees have made it their career goal to be No. 1. It even boasts two second-generation WIVK employees in music director Mickey Dearstone, son of Kenny Dearstone, and on-air personality Jack Ryan, son of operations manager Mike Hammond.

In radio, one would have to look long and hard to find something like that anyplace else or a station with such a rich musical history. WIVK at first followed the old radio formula of having to find musical talent good enough to put on the air in East Tennessee. Today the best talent in the nation finds the station; the combination still works and is the reason why WIVK remains the No. 1 station in East Tennessee.

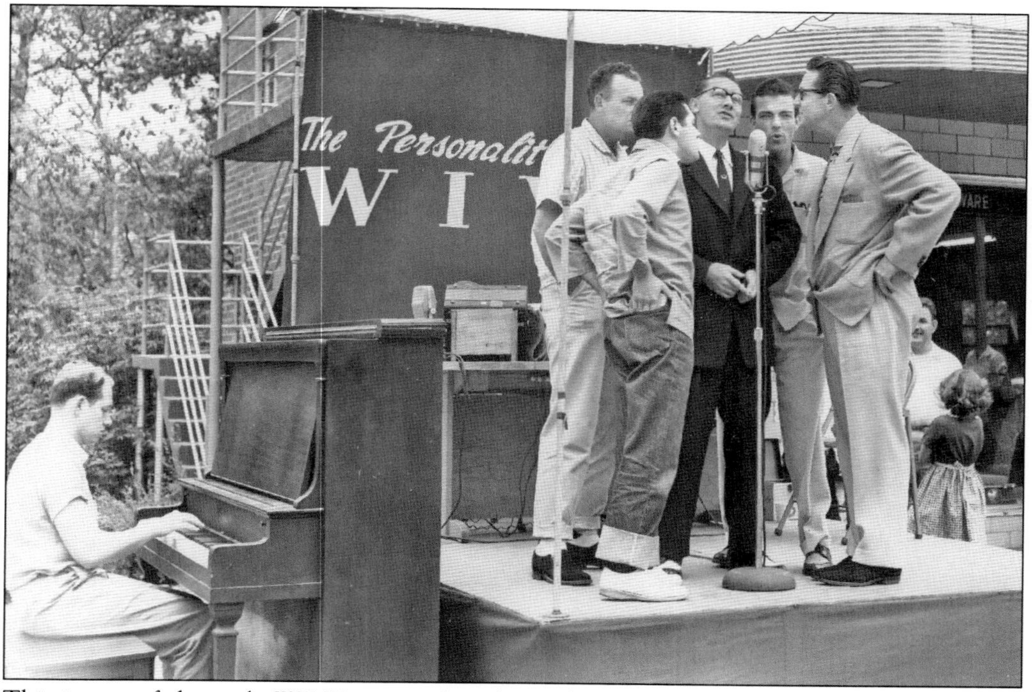

This is one of the early WIVK remote broadcasts featuring a quartet singing with a man at the piano playing. (Photograph courtesy McClung Historical Collection, Knox County Public Library System, Knoxville, Tennessee.)

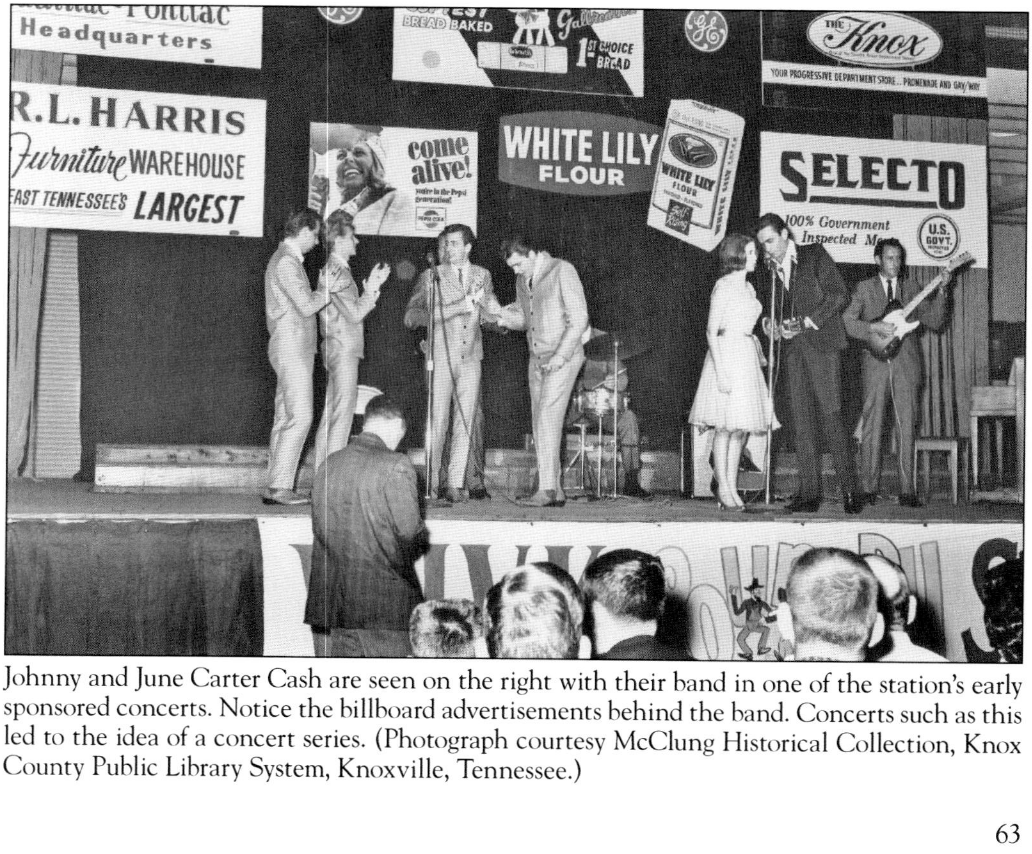

Johnny and June Carter Cash are seen on the right with their band in one of the station's early sponsored concerts. Notice the billboard advertisements behind the band. Concerts such as this led to the idea of a concert series. (Photograph courtesy McClung Historical Collection, Knox County Public Library System, Knoxville, Tennessee.)

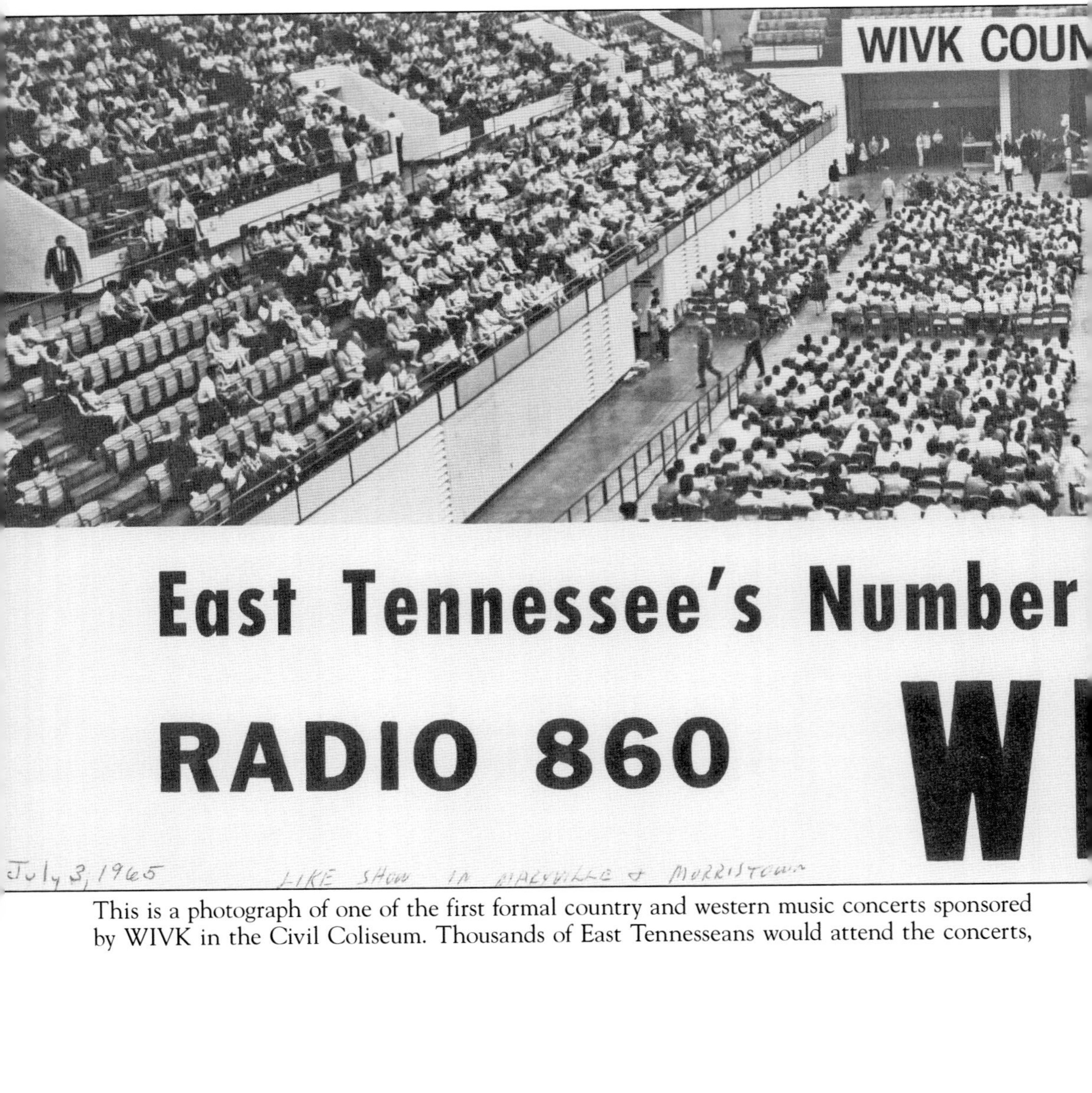

This is a photograph of one of the first formal country and western music concerts sponsored by WIVK in the Civil Coliseum. Thousands of East Tennesseans would attend the concerts,

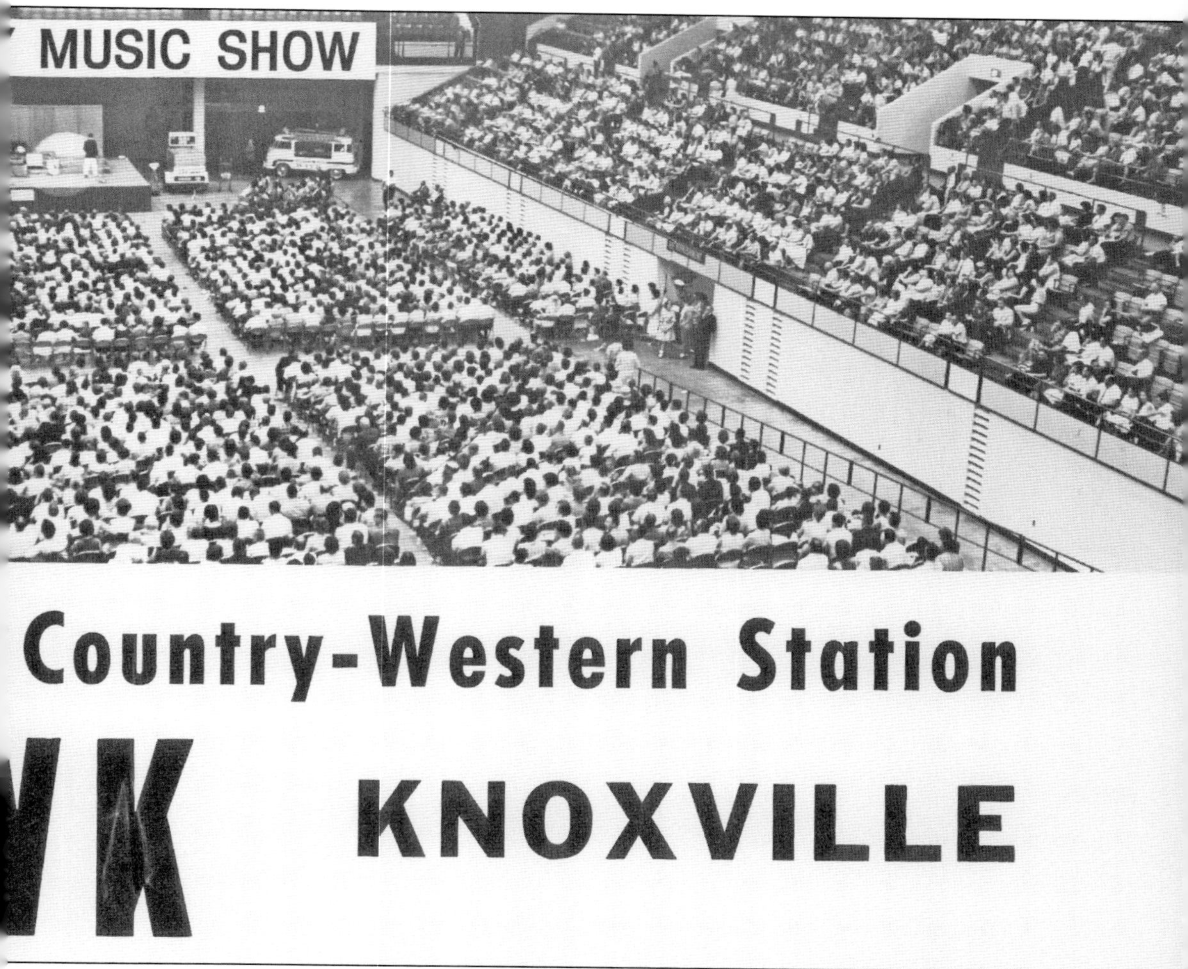

leading to the idea of the WIVK *Shindig.* (Photograph courtesy McClung Historical Collection, Knox County Public Library System, Knoxville, Tennessee.)

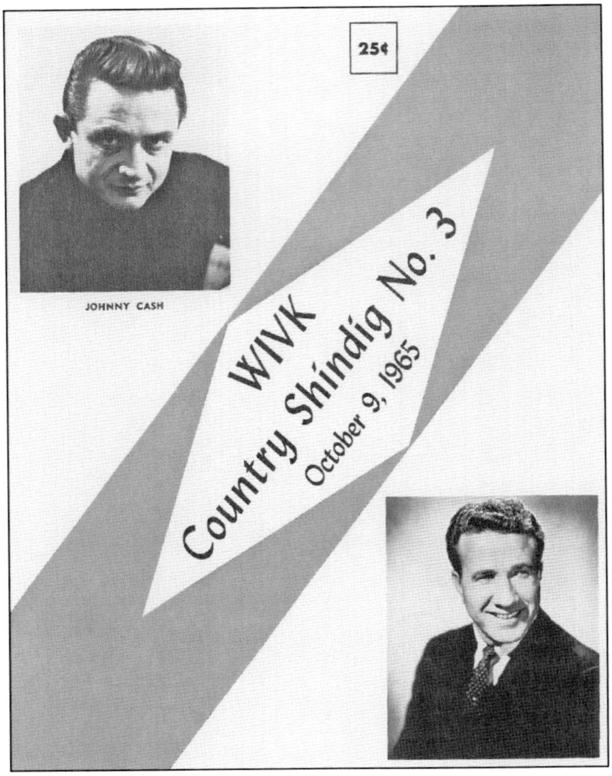

This is the cover of WIVK *Shindig* No. 3, featuring Johnny Cash and popular cowboy singer Marty Robbins. (Photograph courtesy McClung Historical Collection, Knox County Public Library System, Knoxville, Tennessee.)

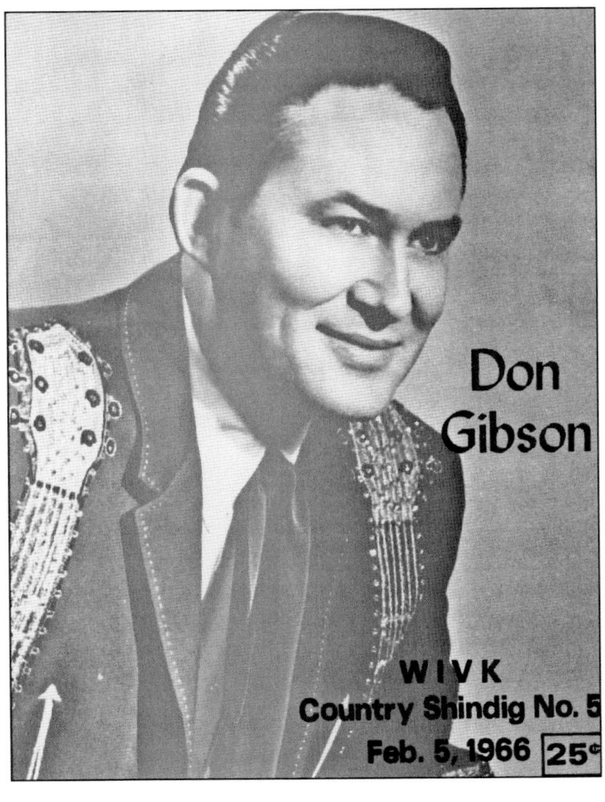

This cover featured Don Gibson. The WIVK *Shindig* programs became collector's items over the years, and few are found today. (Photograph courtesy McClung Historical Collection, Knox County Public Library System, Knoxville, Tennessee.)

The covers of the programs could get interesting at times. This program featured not only the name of the band that was playing, but also listed the names of on-air talent at the station. (Photograph courtesy McClung Historical Collection, Knox County Public Library System, Knoxville, Tennessee.)

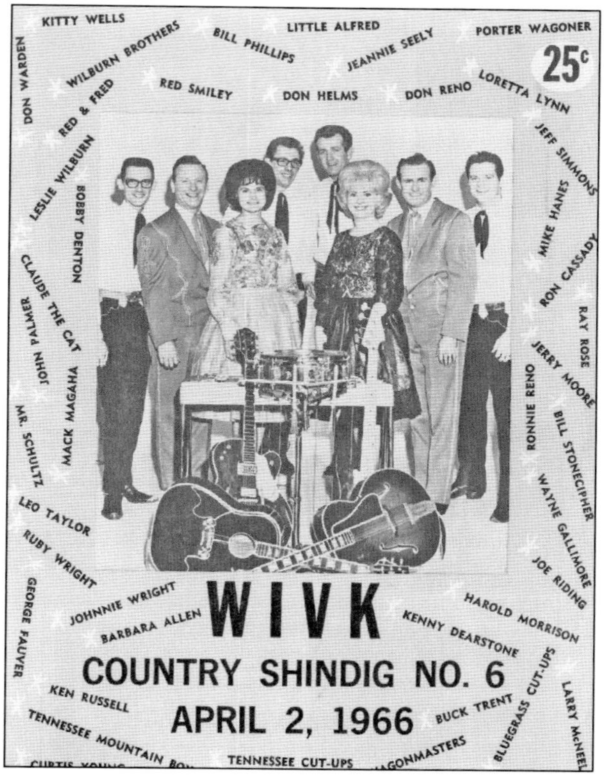

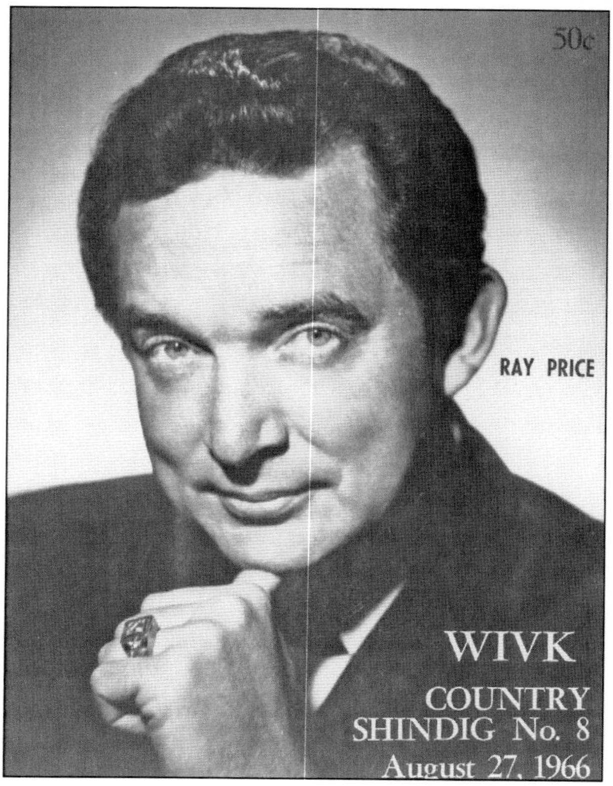

The music of WIVK has evolved over time. Here is a program from *Shindig* No. 8, which featured Ray Price, who was one of America's best-selling artists of the time. (Photograph courtesy McClung Historical Collection, Knox County Public Library System, Knoxville, Tennessee.)

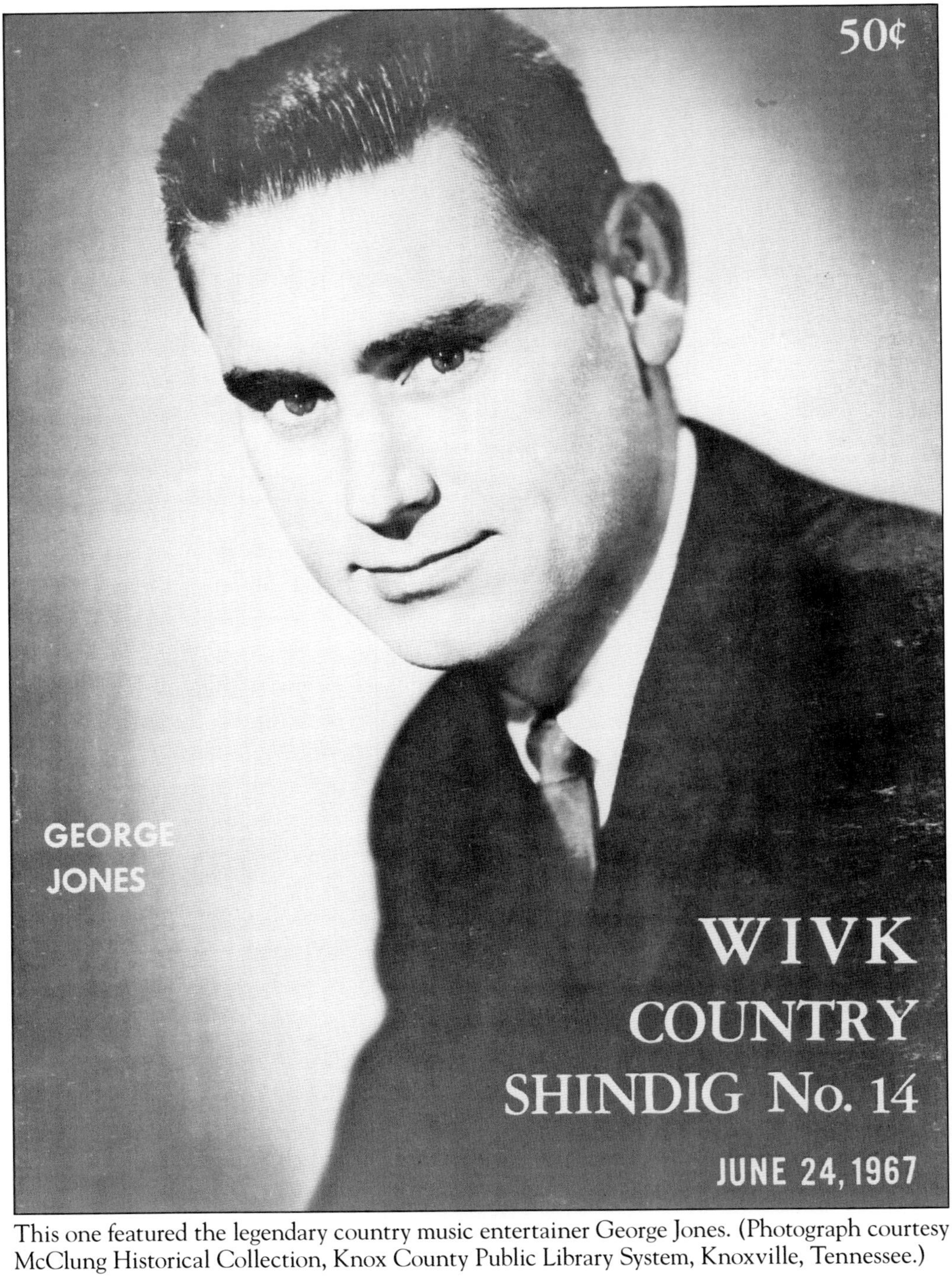

This one featured the legendary country music entertainer George Jones. (Photograph courtesy McClung Historical Collection, Knox County Public Library System, Knoxville, Tennessee.)

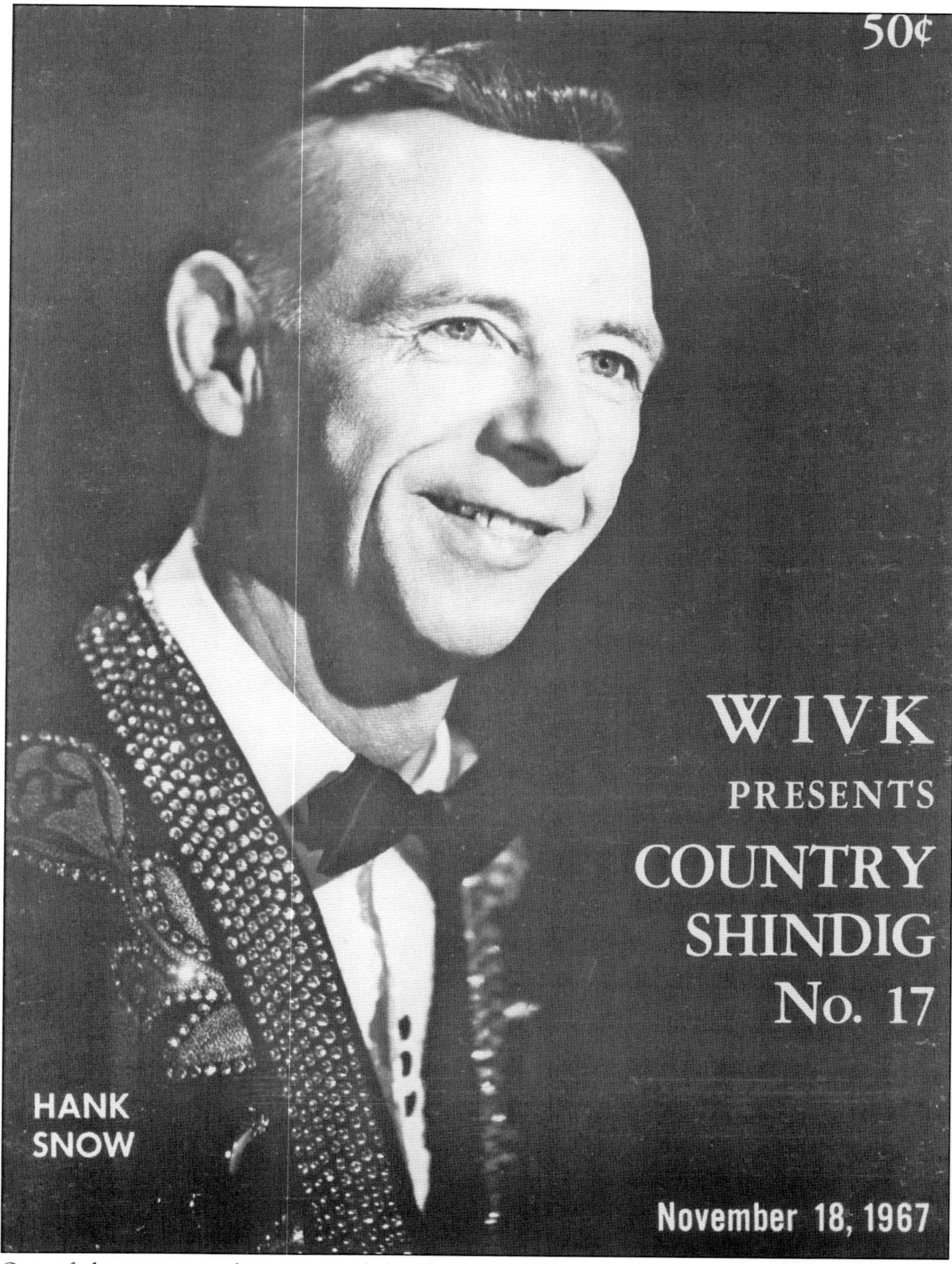

One of the most popular singers of the day, especially with military servicemen, was Hank Snow. The "Singing Ranger" gave birth to Claude Tomlinson's "Hank Range" character. Some of the more memorable moments of the *Shindigs* were the comic relief supplied by WIVK's on-air announcers. (Photograph courtesy McClung Historical Collection, Knox County Public Library System, Knoxville, Tennessee.)

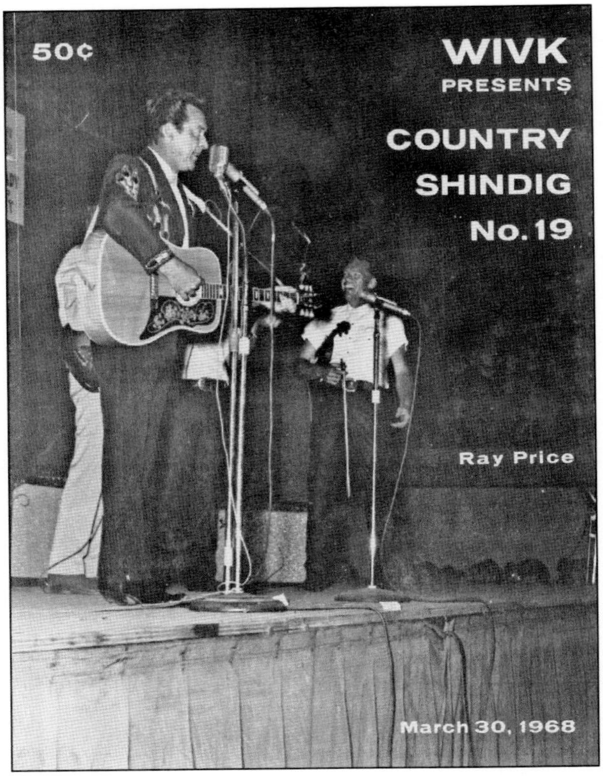

Ray Price was always a popular act in East Tennessee. Here he is featured on the cover of the *Shindig* No. 19 program. The back of the program, seen below, also featured the other acts at the concerts. This one featured country music entertainers Loretta Lynn and Conway Twitty. (Photograph courtesy McClung Historical Collection, Knox County Public Library System, Knoxville, Tennessee.)

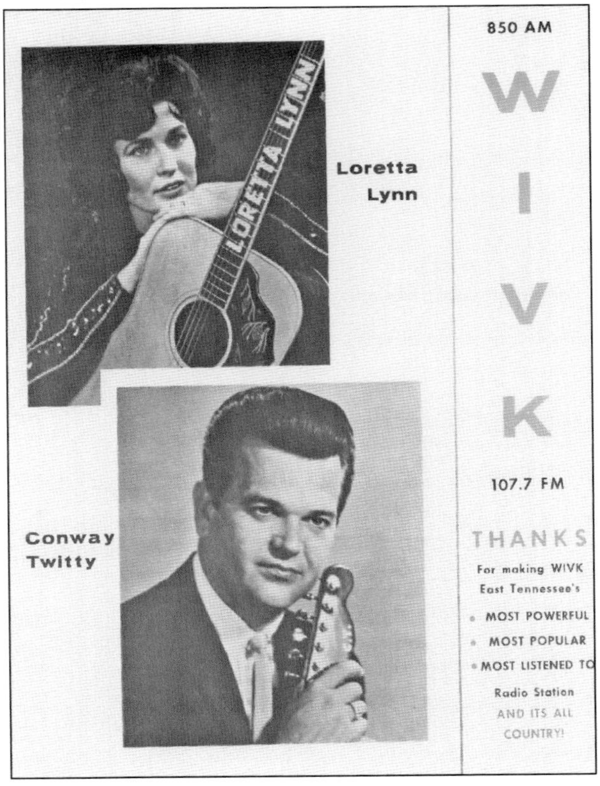

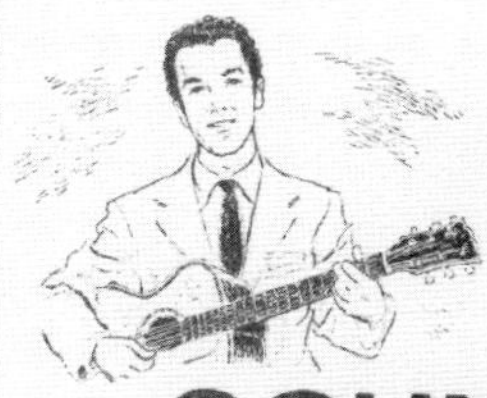

WIVK

PRESENTS

COUNTRY SHINDIG No. 32

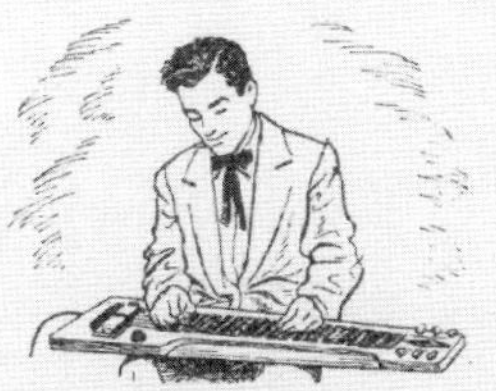

107.7 FM — JANUARY 24, 1970 — 850 AM

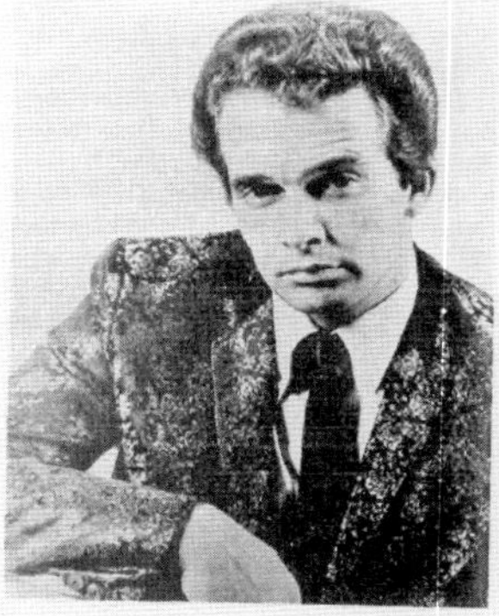

MERLE HAGGARD

Few Country singers have created the interest and excitement with both the public and the trade as has Capitol recording artist, Merle Haggard. From the time he burst into the limelight with his initial recording on Tally Records, "Singing My Heart Out", the singer has built a large and enthusiastic audience. And little wonder, when he continues to turn out such records as his latest release, "The Fightin' Side of Me", a former WIVK Pick Hit, now well on it's way to becoming another Merle Haggard Great. Merle Haggard is a colorful guy, energetic, restless, imaginative, with an abundance of enthusiasm, ideas and talent. You'll be glad you've had this opportunity to meet and hear Merle Haggard and "The Strangers" in person . . . because it's been said time and again: You'll Like 'Em!

PORTER WAGONER AND DOLLY PARTON

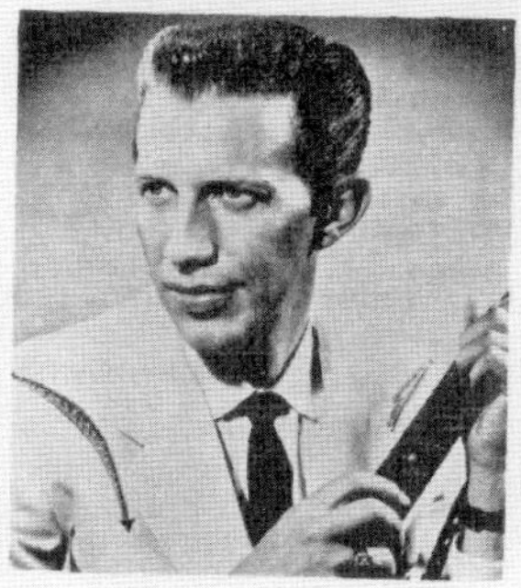

Even if you've never met Porter Wagoner in person, it seems you know him personally. Porter has scores of fans tune in each week to see the highly rated Porter Wagoner Show with Dolly Parton, The Wagonmasters, plus a top name guest each week. Porter always displays his ability to make you feel a part of the show. Country Music is better off today because of this tall, thin gentleman from West Plains, Missouri. From "A Satisfied Mind" in 1955, thru "Skid Row Joe" and "Cold Hard Facts of Life" . . . right on up to "When You're Hot, You're Hot", Porter Wagoner has blasted the charts with Hit after Hit.

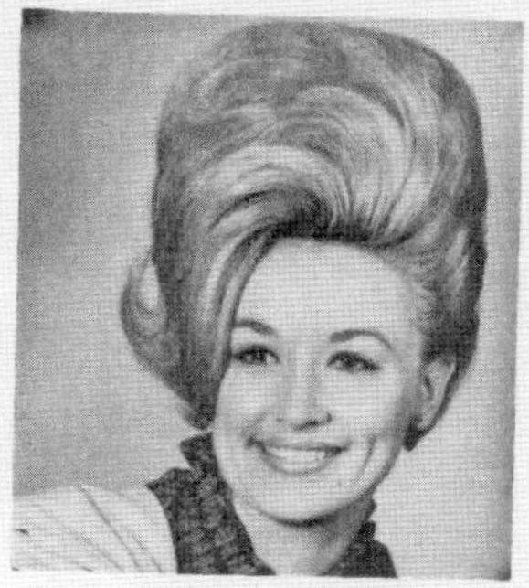

Dolly Parton, a former Knoxville girl and now a regular on the Porter Wagoner Show, has blended her talents with Porter's to re-create an old favorite "Just Someone I Used to Know" . . . and produce a smash country music hit. Dolly's latest, "Daddy Come and Get Me" is already proving to be a favorite with the Dolly Parton fans. Singing and performing alone, or together, Porter Wagoner and Dolly Parton produce a masterful display of talent that's Country Music Entertainment at it's best . . . and, believe me, when they're on-stage, it's a thrill you'll never forget.

DON'T MISS SHINDIG NO. 33

March 14, 1970

By 1970, the five-year concert series by WIVK had evolved, as did the programs. This was *Shindig* No. 32, with Merle Haggard, Porter Waggoner, and local talent Dolly Parton. The biographies about the performers were a new twist for the station. (Photograph courtesy McClung Historical Collection, Knox County Public Library System, Knoxville, Tennessee.)

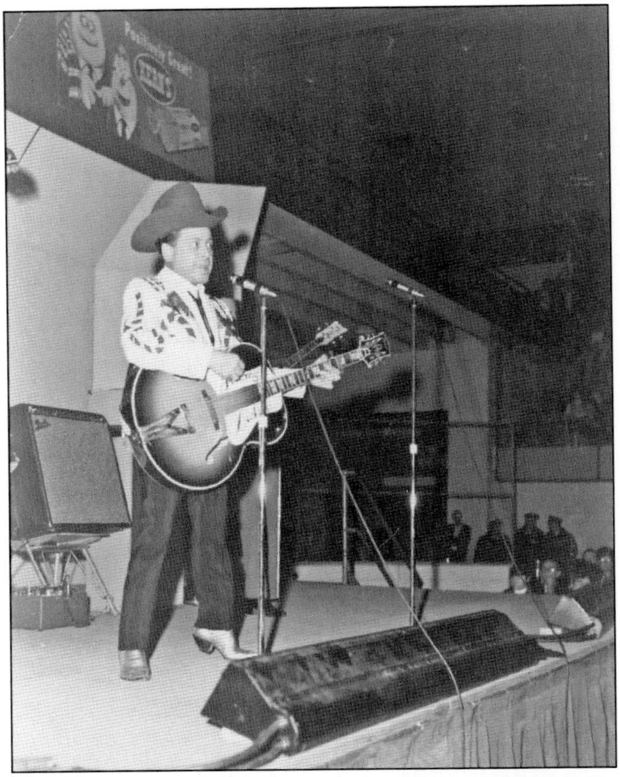

This WIVK concert performance was by Country Music Hall of Famer "Little" Jimmy Dickens, who made a number of Knoxville appearances on behalf of the station. The Grand Ole Opry star was a popular entertainer in East Tennessee. (Photograph courtesy McClung Historical Collection, Knox County Public Library System, Knoxville, Tennessee.)

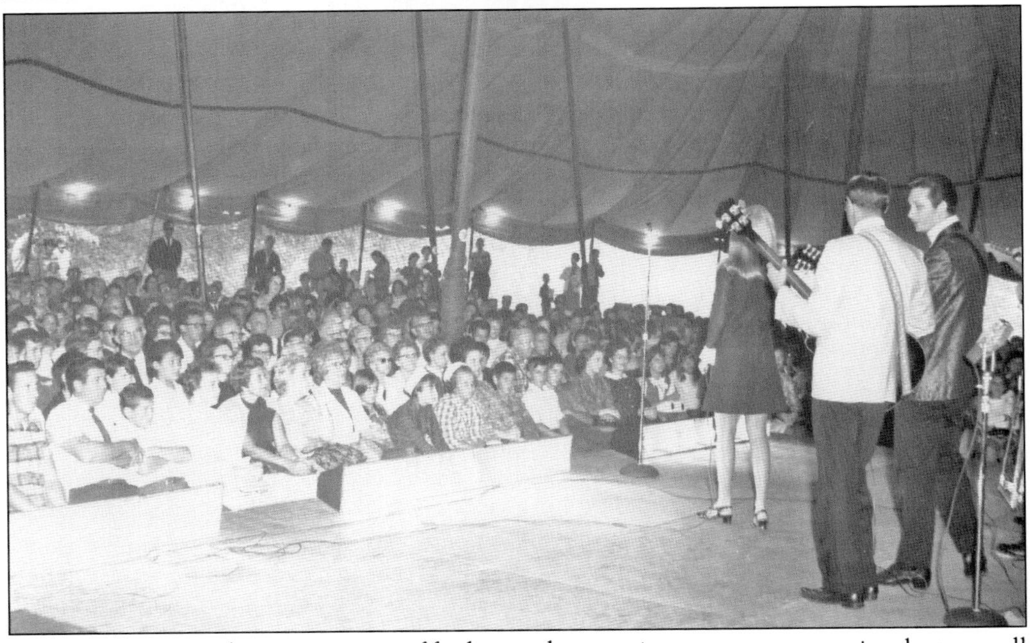

WIVK also remained true to its roots of finding and supporting new country music talent, as well as holding free concerts as a listener appreciation promotion. One of the most popular places for the station to do this was at the annual TVA&I Fair in Chilhowee Park. Here the Kountry Kings with Kathy Hill perform for an audience. (Photograph courtesy McClung Historical Collection, Knox County Public Library System, Knoxville, Tennessee.)

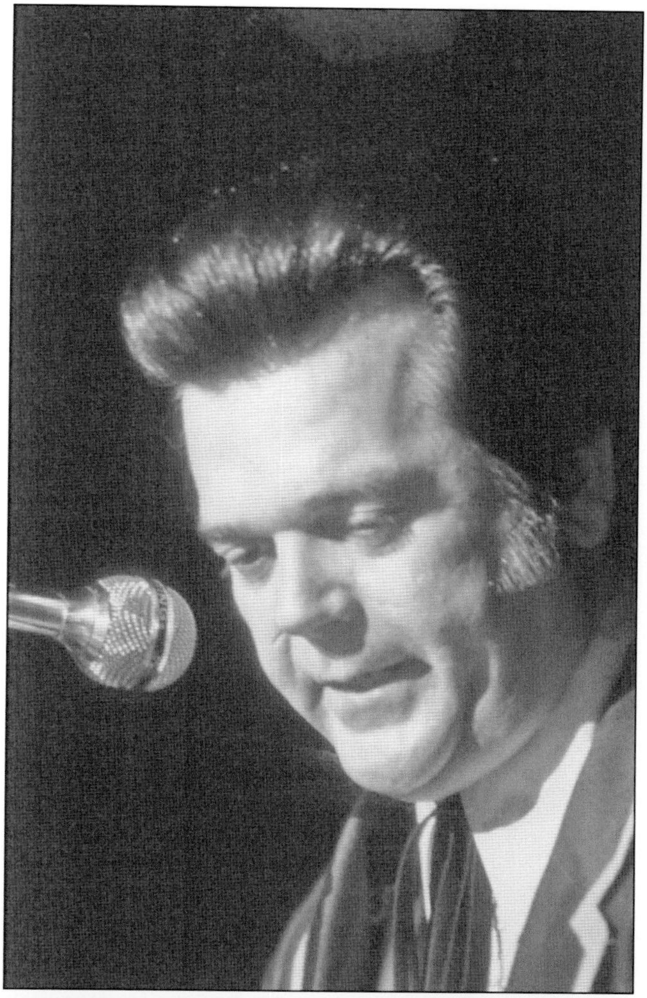

The image at right is of Conway Twitty in concert at a WIVK *Shindig.* Pictured below are Mel Tillis and his band. The announcers at the concert always had fun with Tillis and his stuttering voice. It was always in good fun, and he became a close friend of the station.

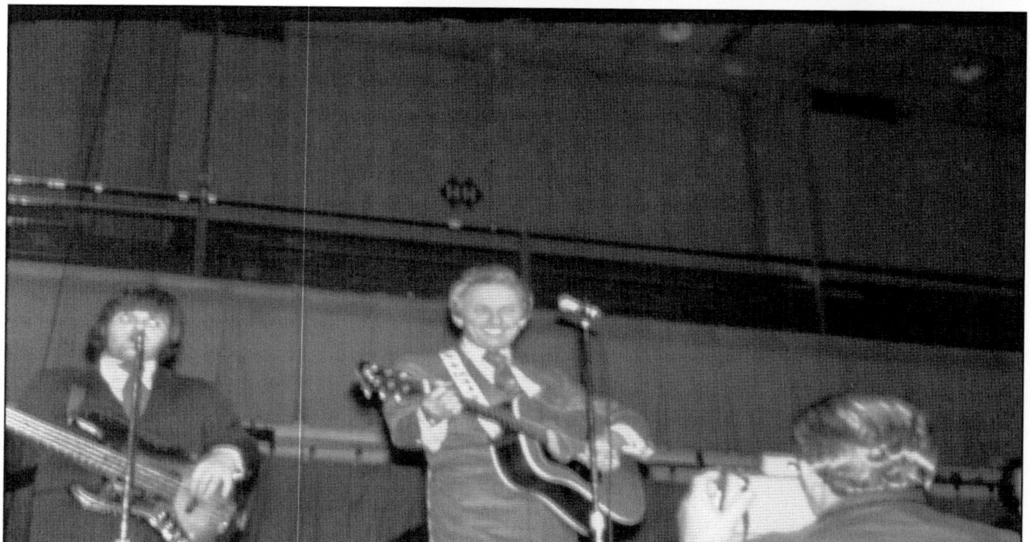

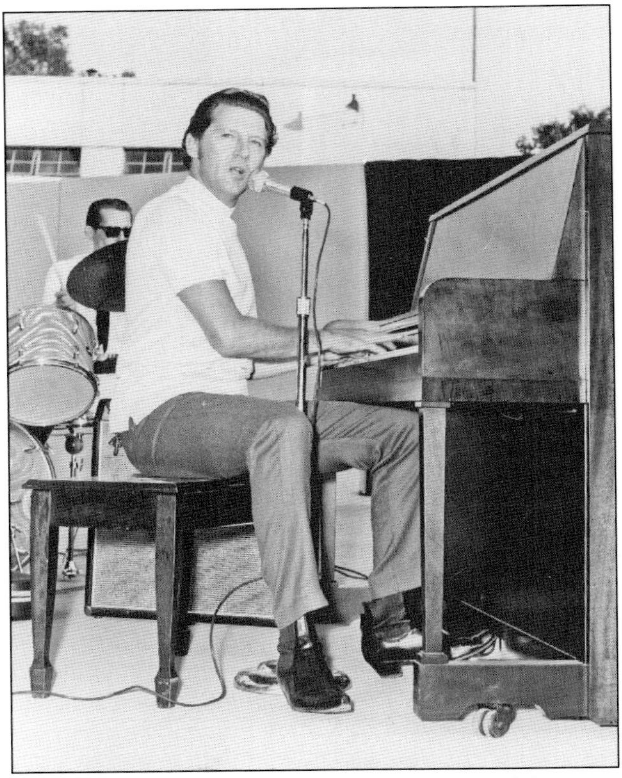

Jerry Lee Lewis was also a popular *Shindig* performer who could work the crowd like no other entertainer. Here he is seen at the piano in an early performance hosted by the station. (Photograph courtesy WIVK.)

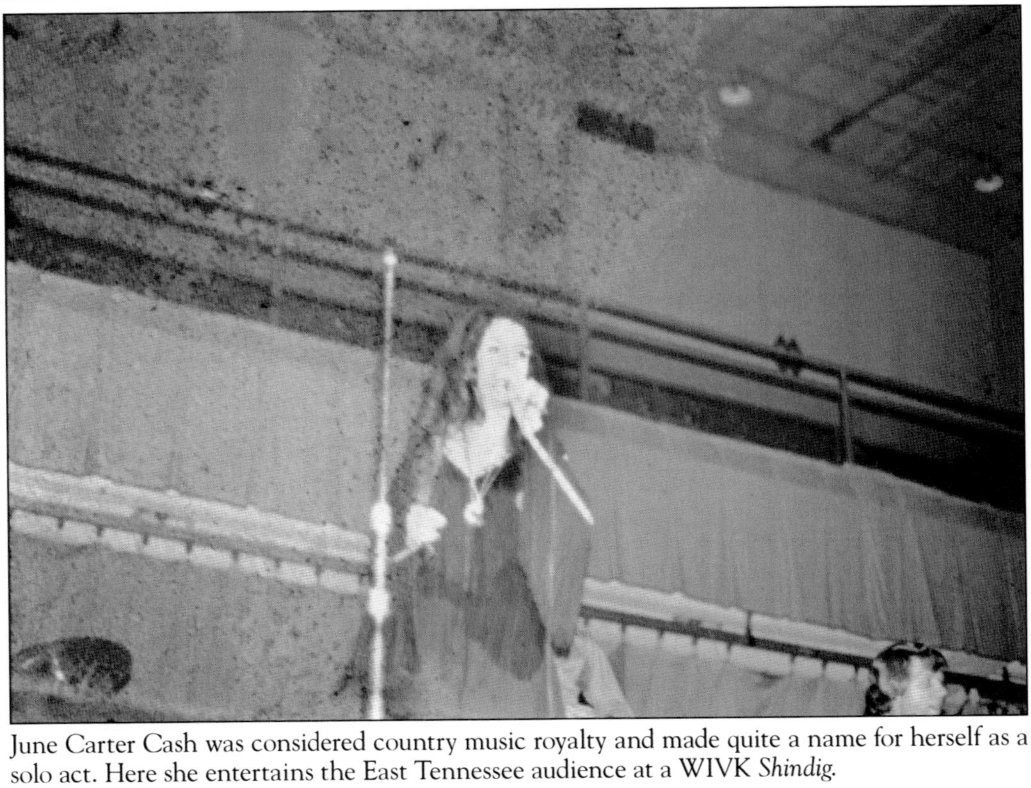

June Carter Cash was considered country music royalty and made quite a name for herself as a solo act. Here she entertains the East Tennessee audience at a WIVK *Shindig*.

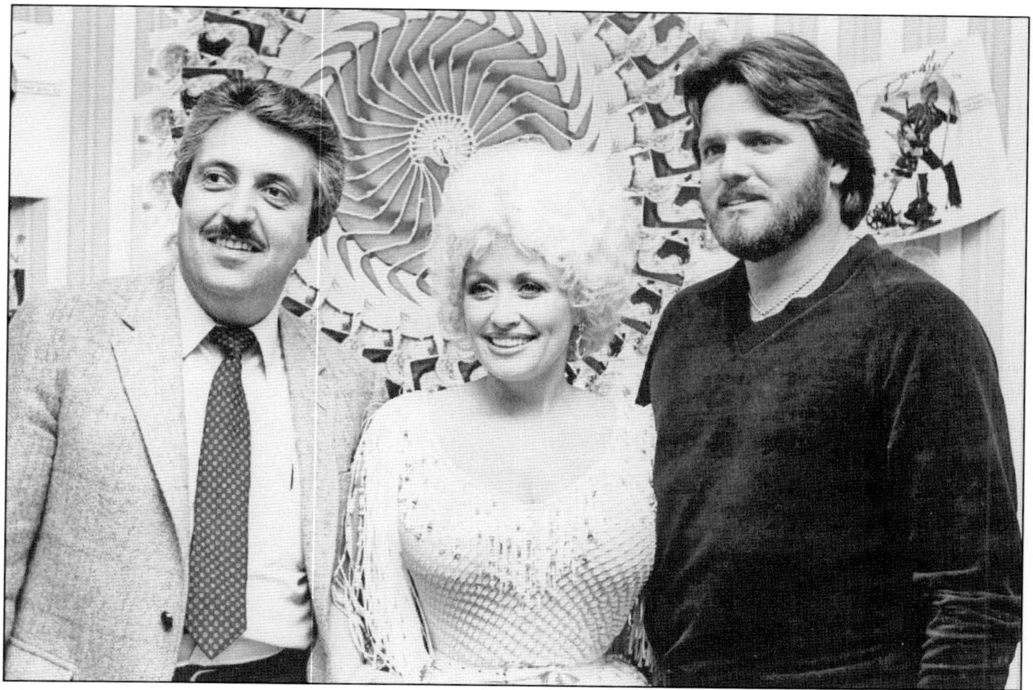

If ever a country music entertainer stayed true to her hometown, it was Dolly Parton. Here she is pictured with local country music artist Con Hunley (right) and WIVK general manager Bobby Denton. (Photograph courtesy WIVK.)

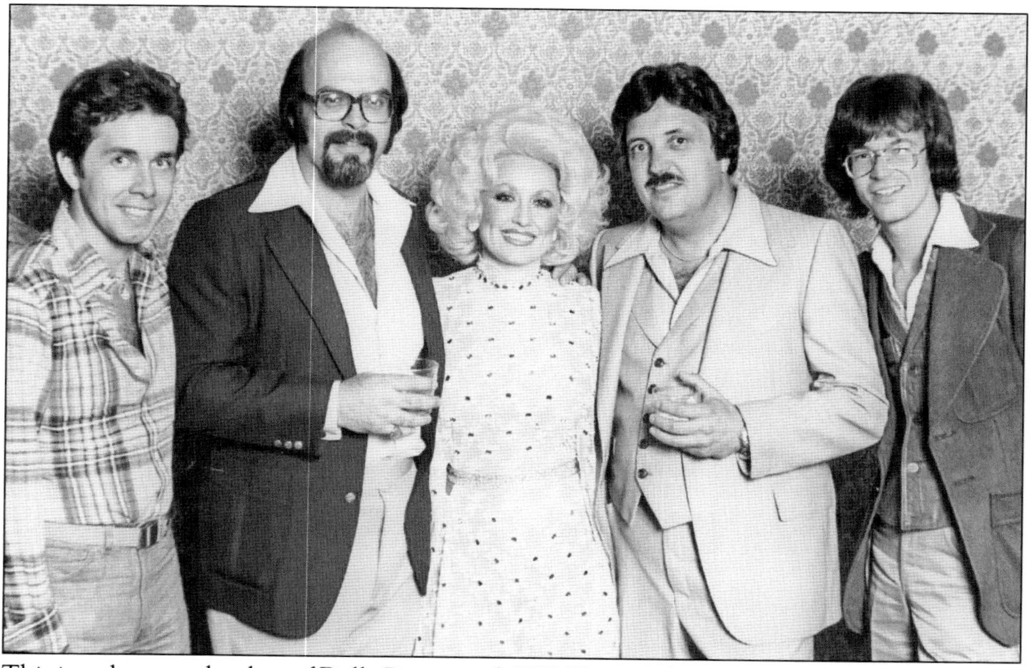

This is a photograph taken of Dolly Parton with WIVK general manager Bobby Denton (second from right), Nashville sister station general manager Bill Hayes (second from left), and record label executives. She is still a frequent visitor at the station and is considered family. (Photograph courtesy WIVK.)

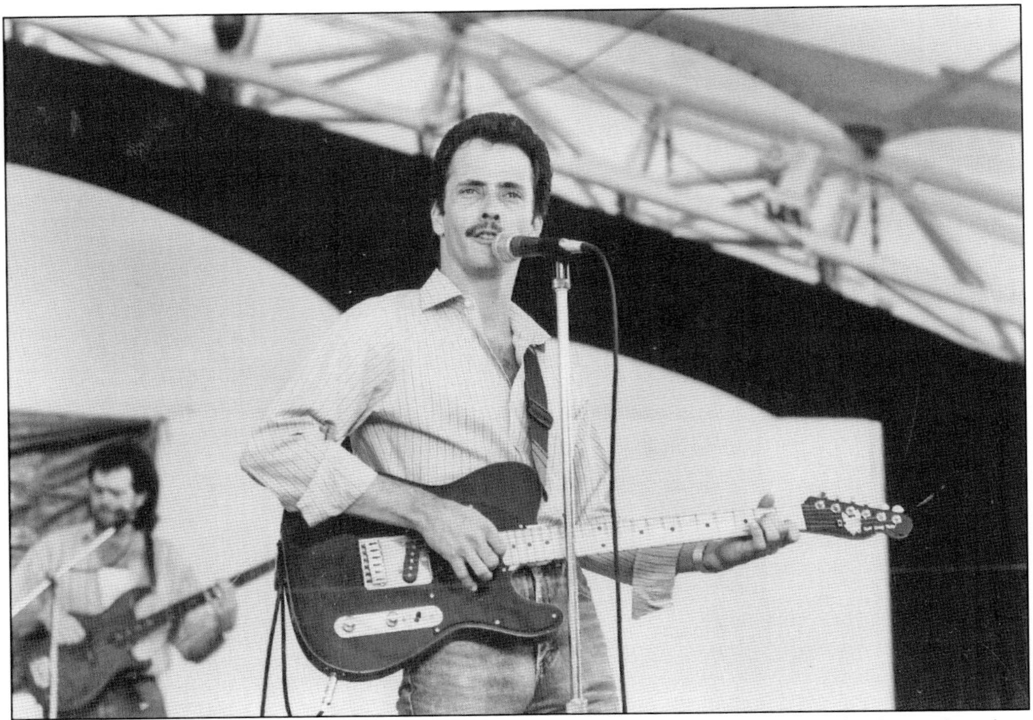

The Chuck Wagon Gang was another popular act in East Tennessee and often played at WIVK *Shindigs*.

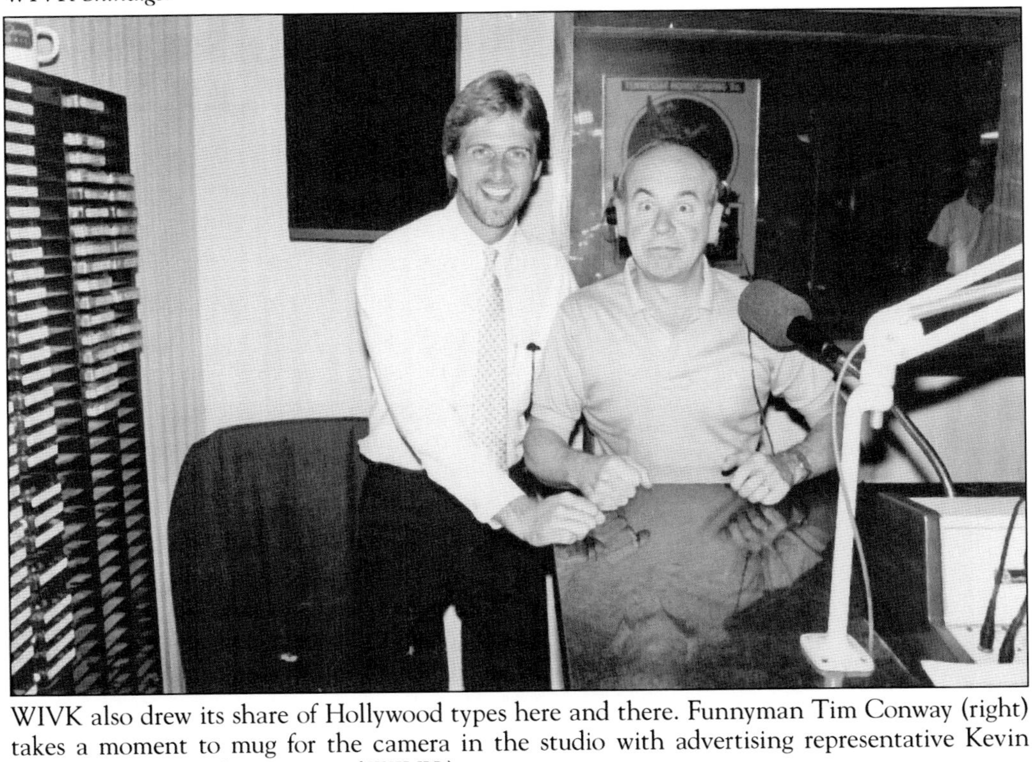

WIVK also drew its share of Hollywood types here and there. Funnyman Tim Conway (right) takes a moment to mug for the camera in the studio with advertising representative Kevin McGee. (Photograph courtesy of WIVK.)

WIVK moved its concert series outside sometimes to Red Gate farms in nearby Union County. Here Con Hunley prepares to perform for the outdoor crowd. Behind him on the right is East Tennessee banker C. H. Butcher. (Photograph courtesy of Ed Brantley.)

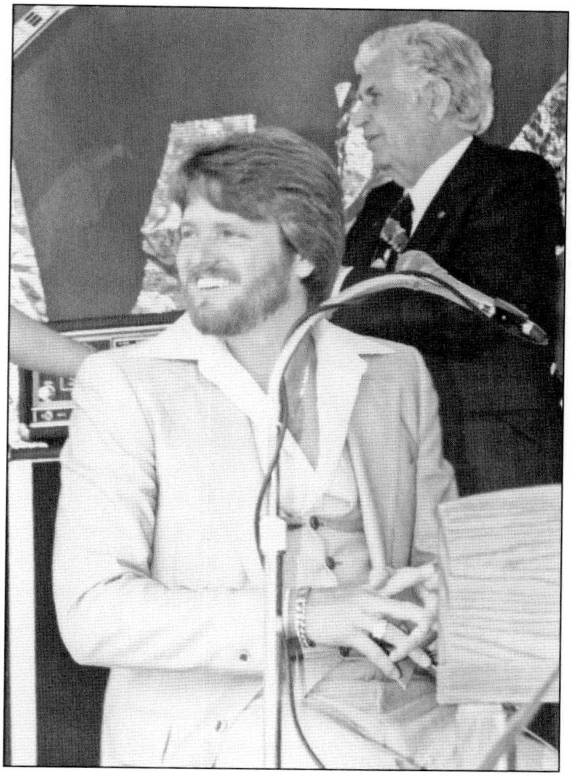

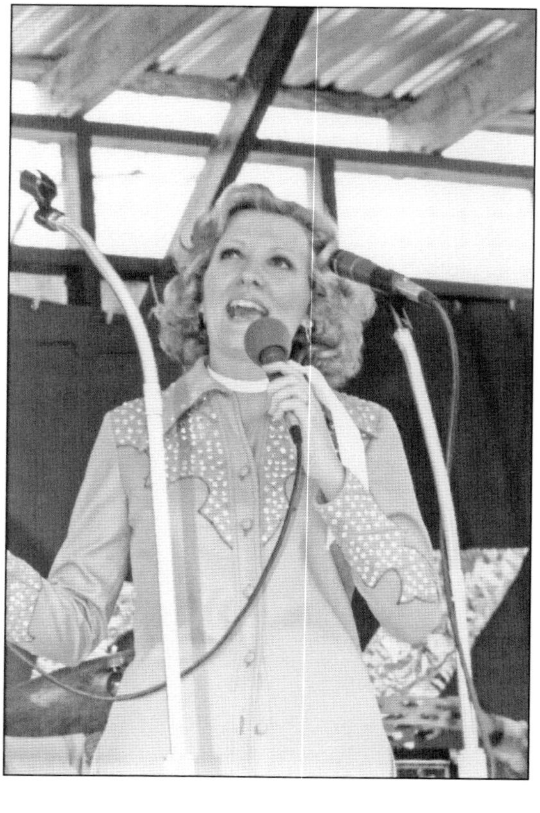

Another homegrown talent who took to the stages at Red Gate was country music singer Ava Barber, who was a frequent guest on the long-running *Lawrence Welk Show*. (Photograph courtesy of Ed Brantley.)

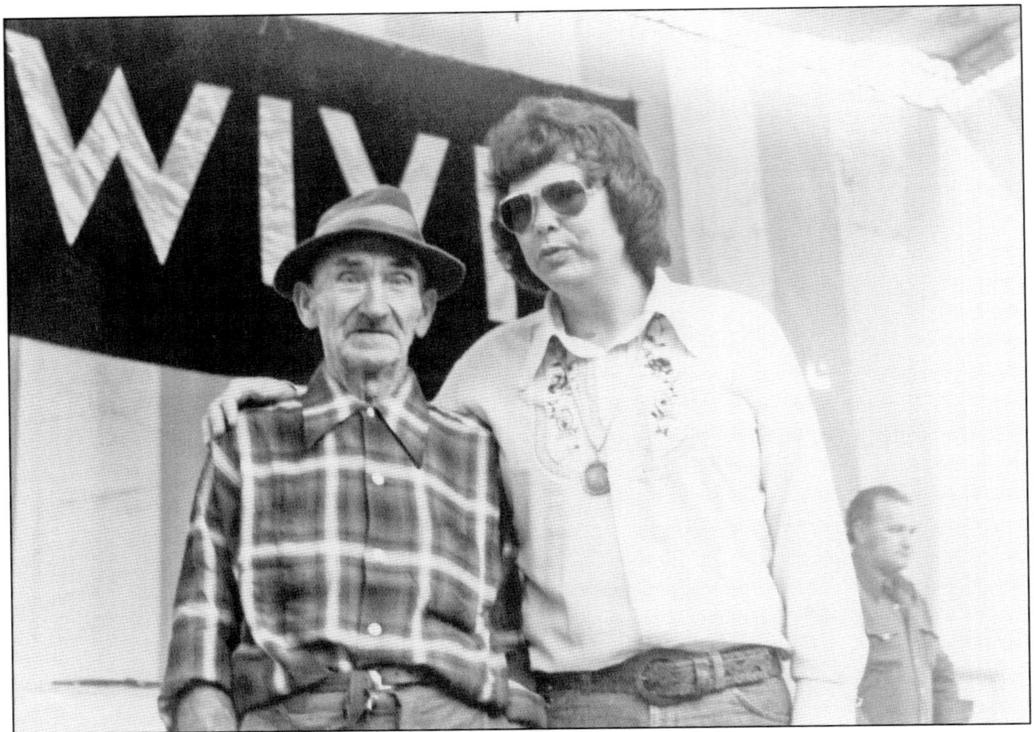

Country music singer Ronnie Milsap, who grew up over the Smoky Mountains in Robbinsville, North Carolina, was another singer who was part of the WIVK family. Here he stops to take a photograph with his grandfather, who had come to hear him perform. (Photograph courtesy Ed Brantley.)

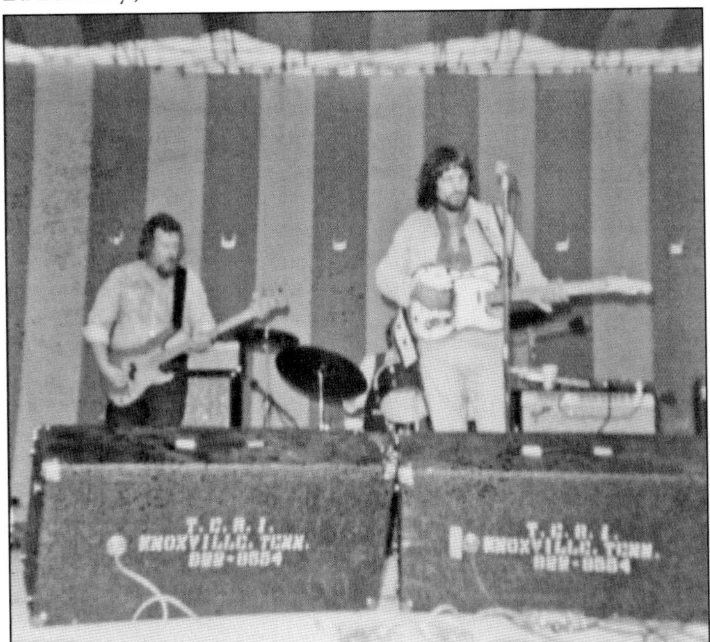

Legendary country music entertainer Waylon Jennings (right) would pack in crowds from all over the region during the "Outlaw" phase of country music in the 1980s. (Photograph courtesy Ed Brantley.)

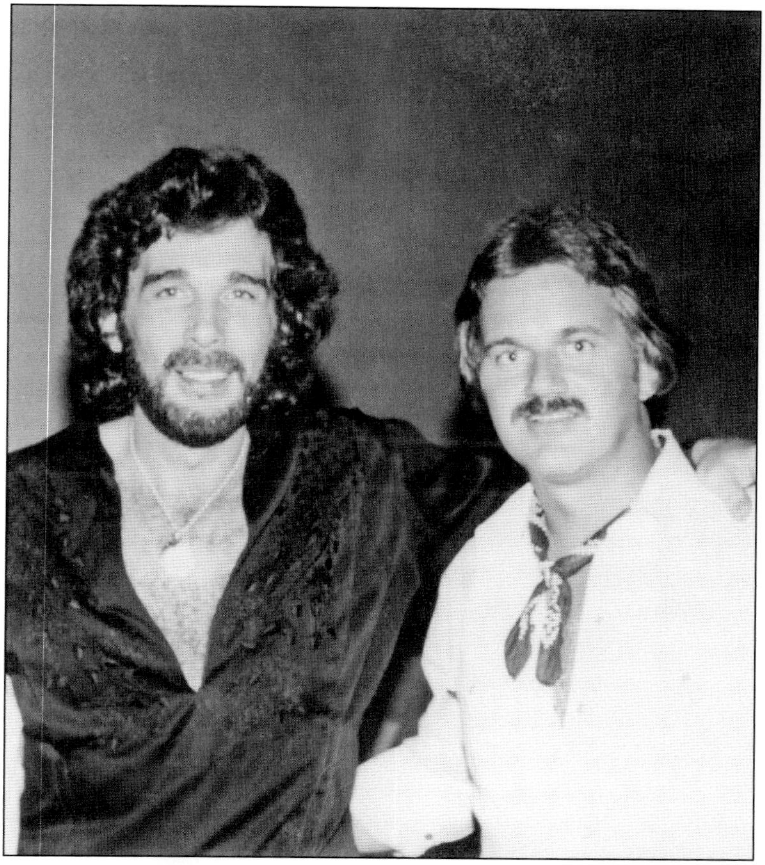

Ed Brantley (left) and Eddie Rabbitt took time for this photograph in the 1980s. Below, WIVK staff share a joke in this photograph with country music entertainer Lee Greenwood (second from left). Mickey Dearstone is second from right. (Photograph courtesy Ed Brantley.)

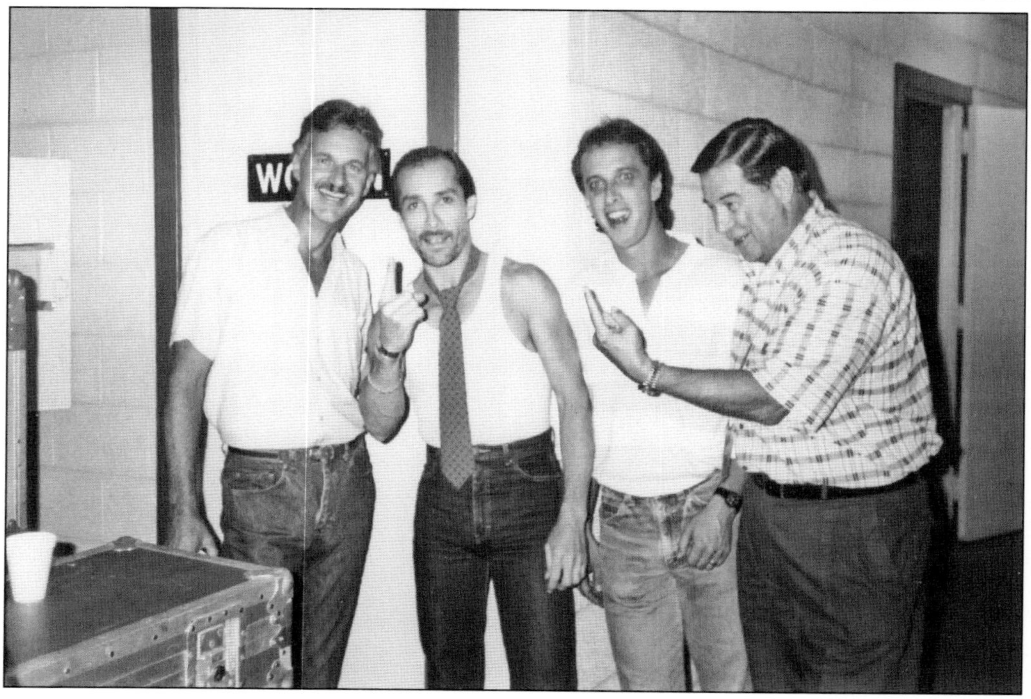

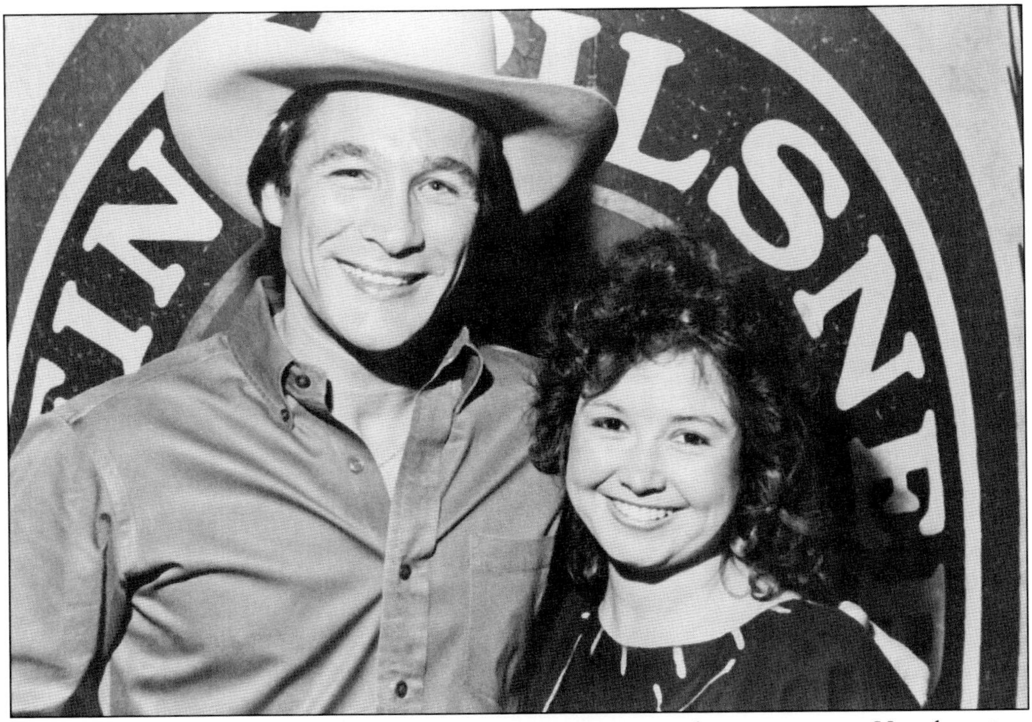

Listener appreciation parties were another way WIVK promoted country music. Here longtime station employee and current traffic manager "Little" Jan Oliver, who was Ethel of the "Ed and Ethel" segments, and country music entertainer Clint Black pose for a photograph. (Photograph courtesy Jan Oliver.)

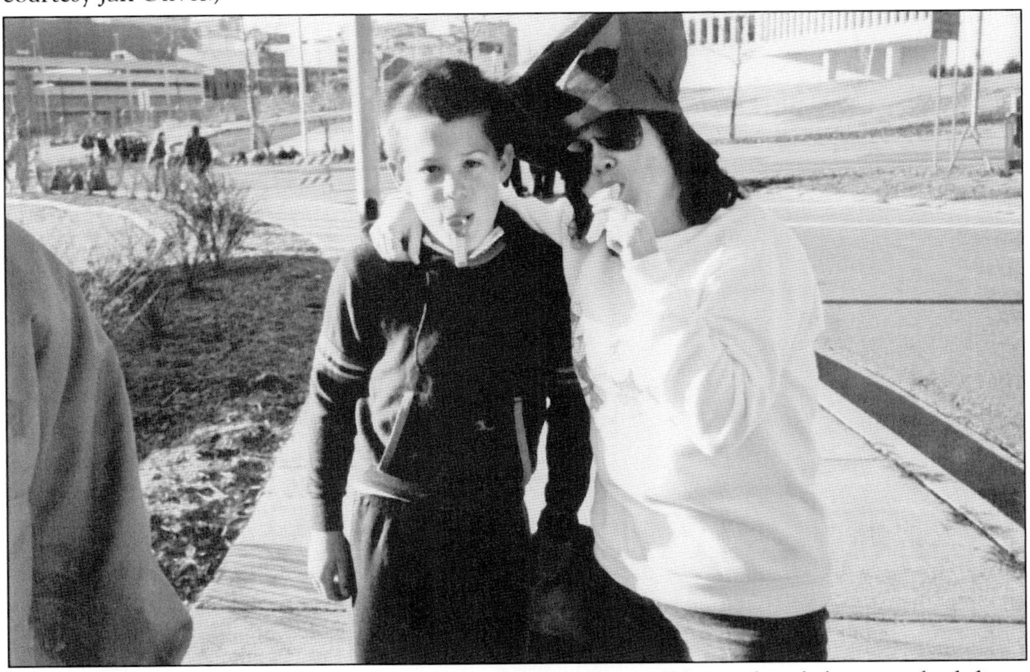

Oliver is getting a young Jack Ryan ready for Ed Brantley's all-kazoo band that marched down Gay Street. Ryan is now the nighttime announcer for WIVK. (Photograph courtesy WIVK.)

As country music changed and truly became the music of America, so did some of the faces that had once been in popular music. Here Marie Osmond sings before a crowd at a WIVK-sponsored concert in Knoxville. (Photograph courtesy WIVK.)

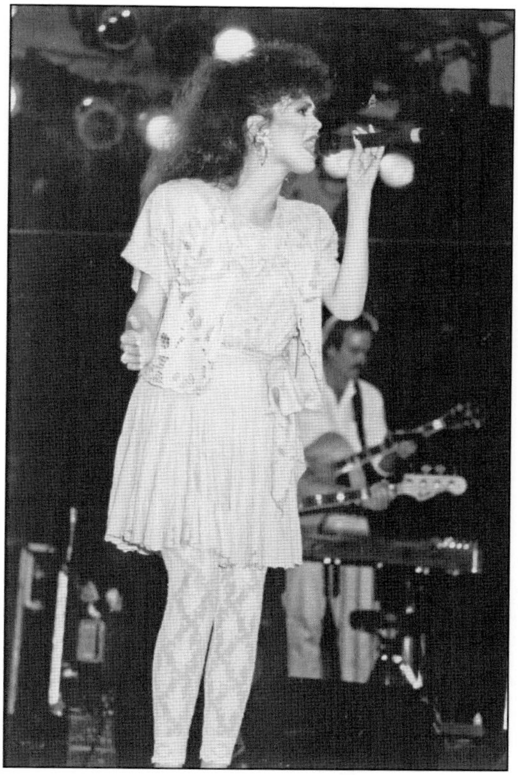

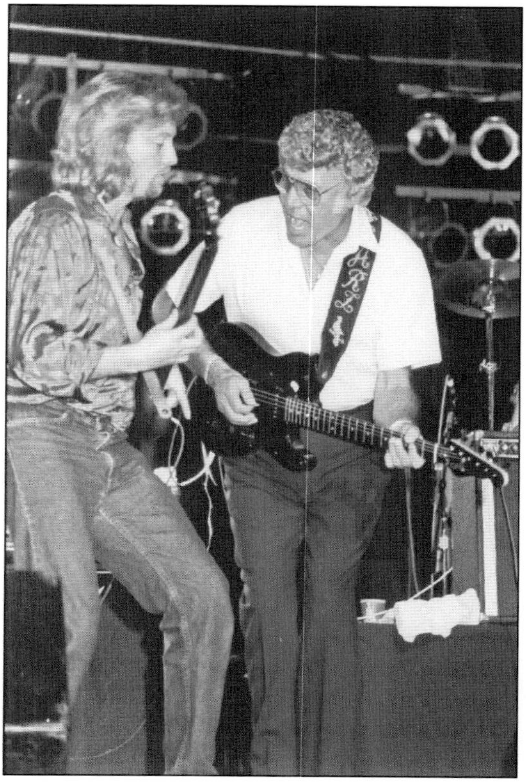

Blending the old with the new became fashionable in country music as some of the old stars were rediscovered by a new generation of listeners at WIVK. Here Carl Perkins (right) brings his brand of music on stage in Knoxville. (Photograph courtesy WIVK.)

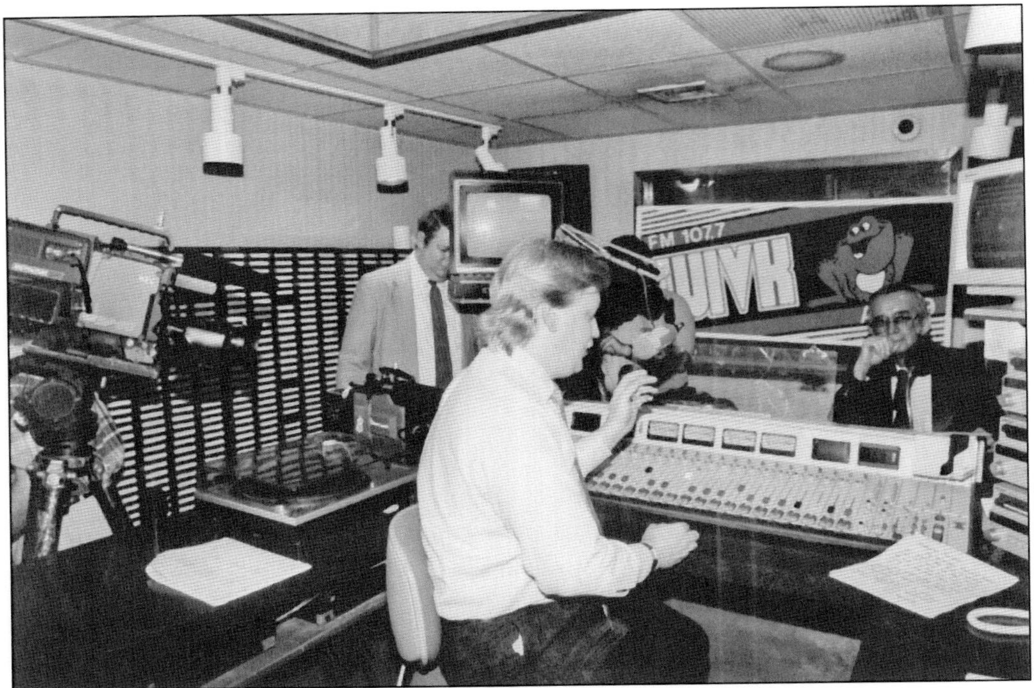

The musical legends coming and going through WIVK made it an interesting place for local news crews. Here Claude Tomlinson (center) and producer Jim Donovan (far right) are being filmed by a local television station. (Photograph courtesy of WIVK.)

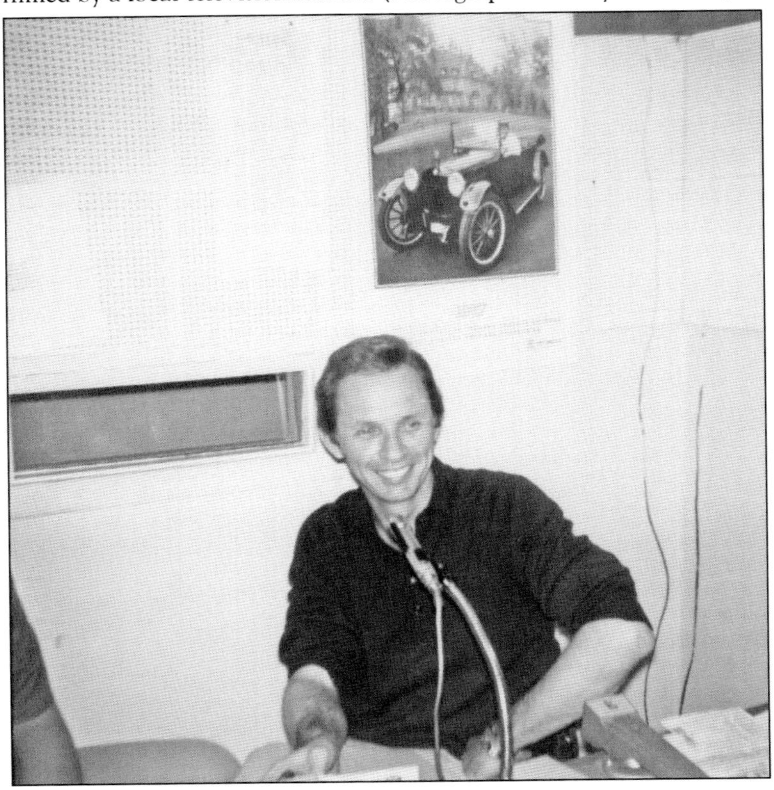

Country music entertainer Mel Tillis laughs at a joke. He was often seen at the station and stopped by WIVK to help celebrate its 30th anniversary.

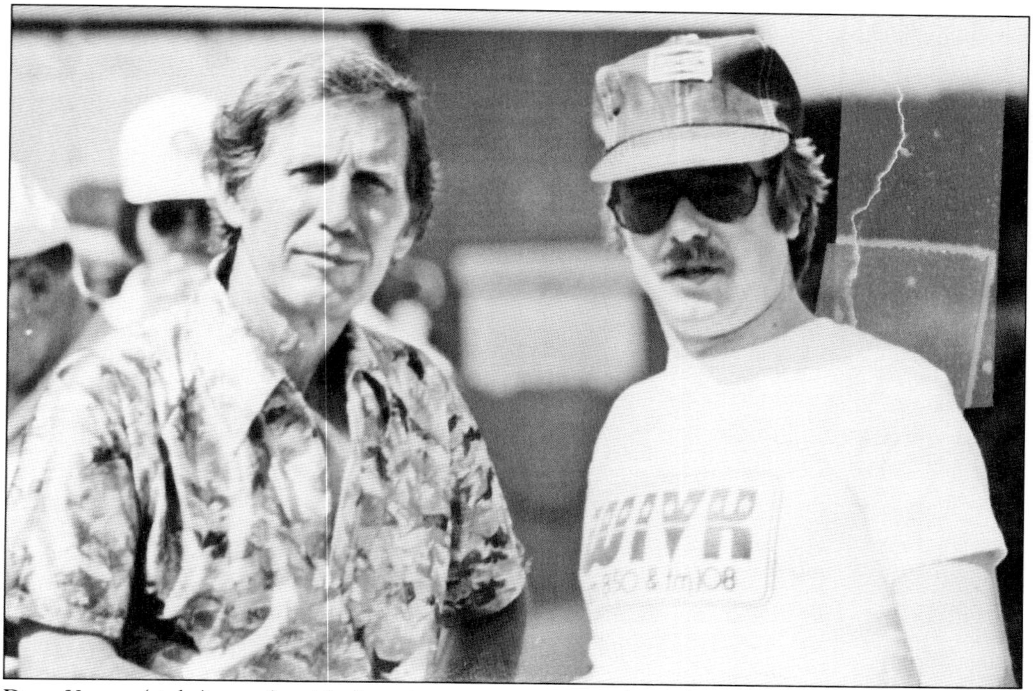

Dave Young (right) stands with American country music legend Chet Atkins, who had serious roots in Knoxville. Atkins once taught guitar lessons on Gay Street near the original WIVK studio. He was later hired to play background music for commercials on competing station WNOX before moving on to Nashville.

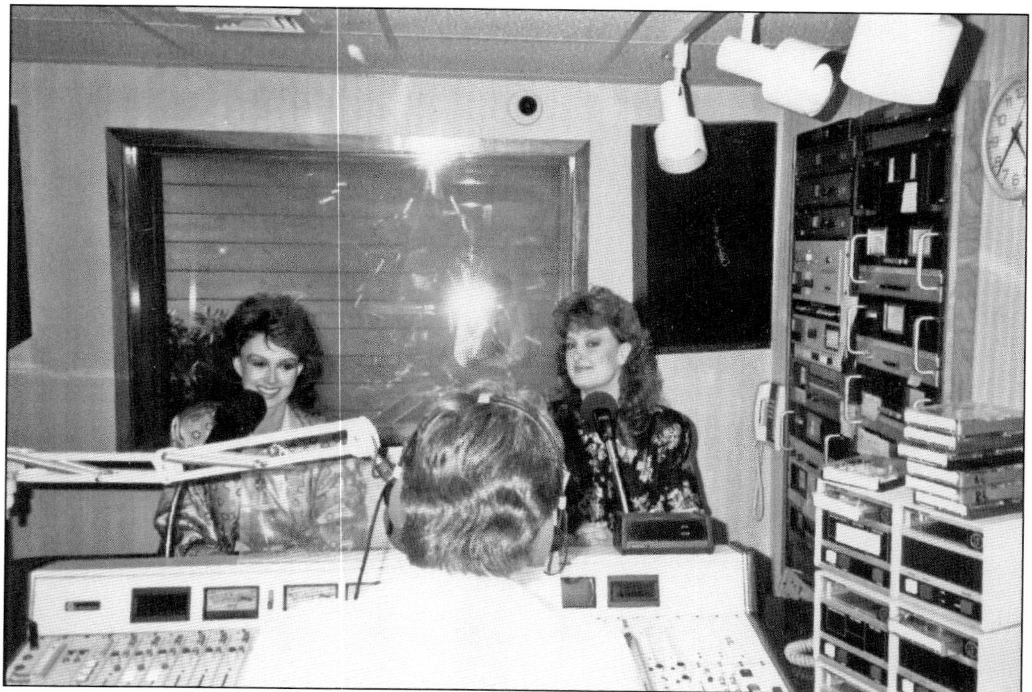

Here Naomi (left) and Wynonna Judd take a few moments in the WIVK studio to answer questions and promote their WIVK concert. (Photograph courtesy WIVK.)

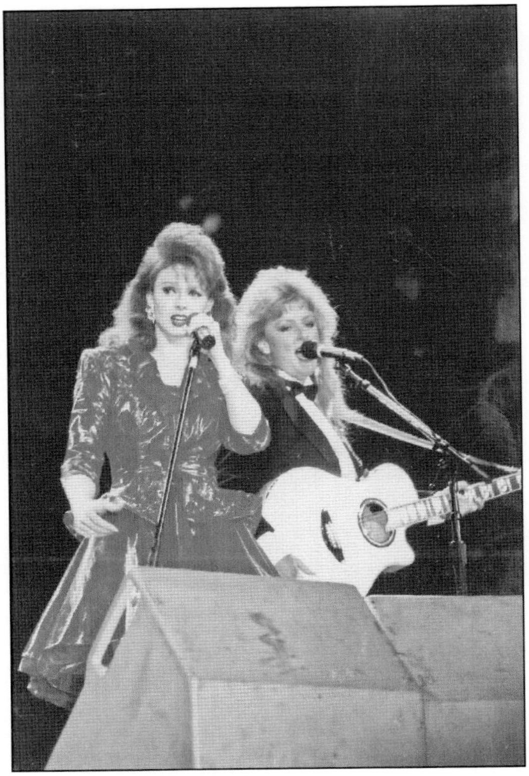

The popular mother-daughter Judds duo made a number of appearances in the city during the 1980s and were among some of the most popular bands WIVK helped promote early in their career. Below, Reba McEntire was also one of the more popular singers of the 1980s and 1990s, and performed in a number of concerts in conjunction with the station. (Photographs courtesy WIVK.)

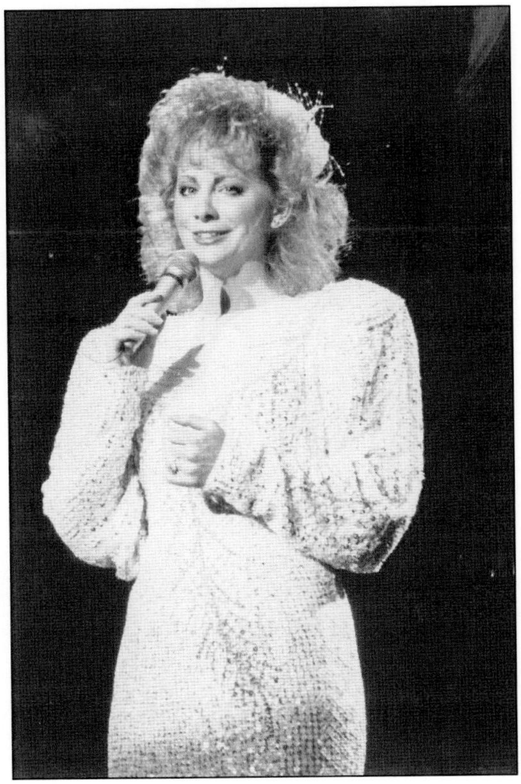

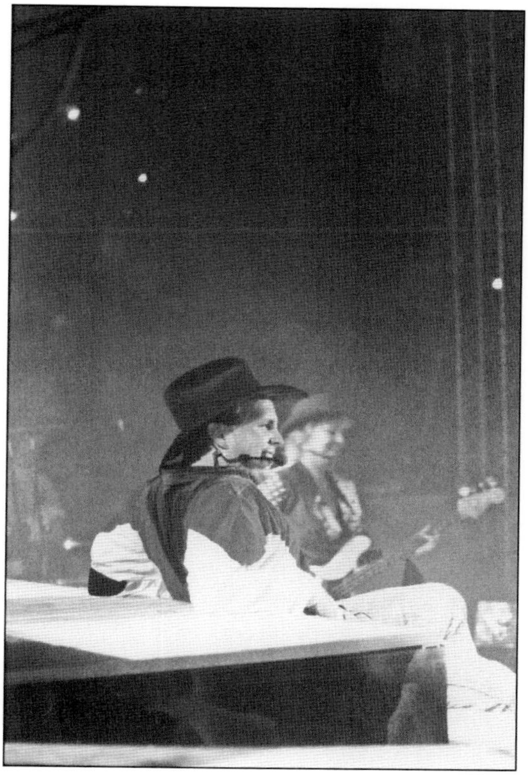

One of the megastar country music entertainers whose career was helped by WIVK was singer Garth Brooks. His different style of music won over an entirely new generation of country music fans and listeners for the station. At right, he takes a break for a moment at a station-promoted concert. Below, Garth Brooks is in full performance mode. Early on in his career, he often stopped by WIVK to visit with announcers and help promote his new albums. (Photographs courtesy WIVK.)

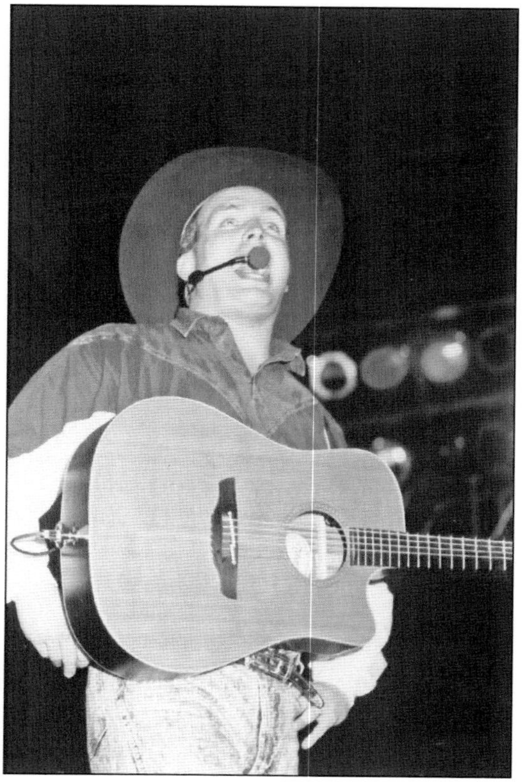

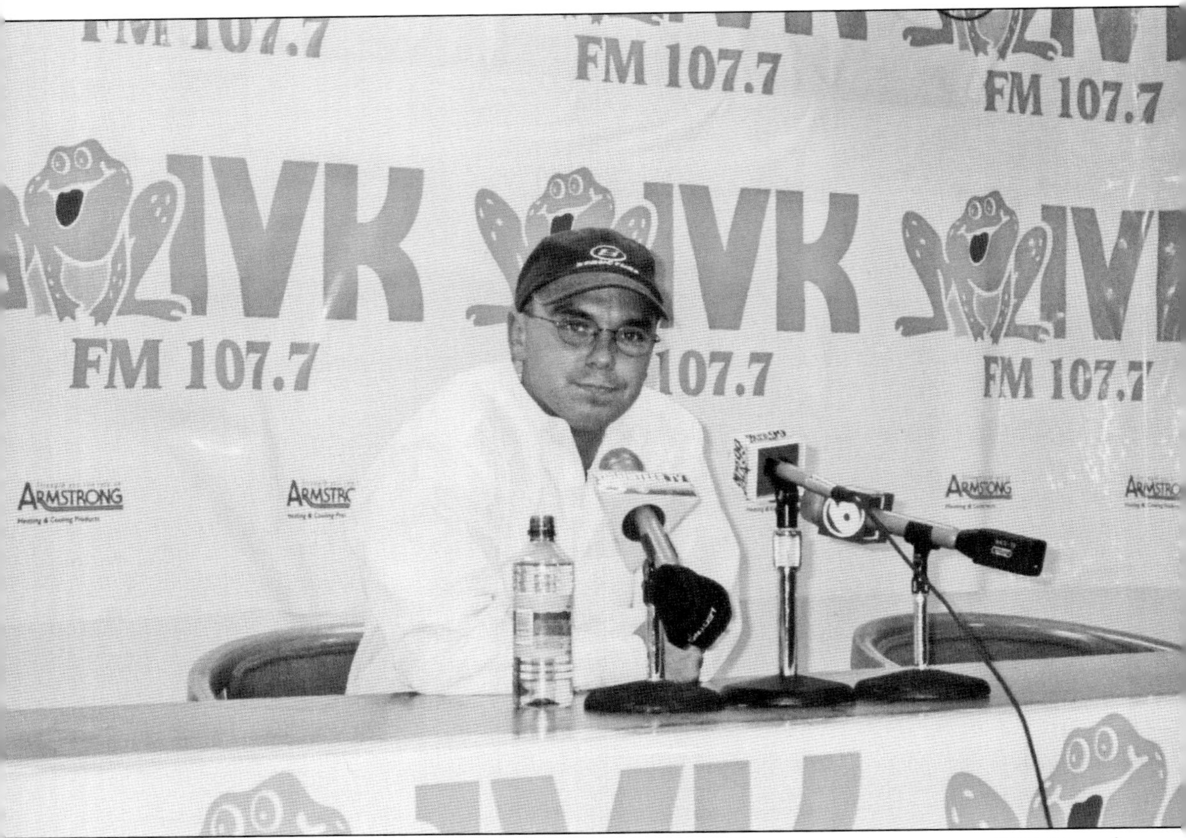

Another Knoxville talent and Gibbs High School alum who grew up listening to WIVK and went on to become a country superstar in his own right is singer Kenny Chesney. Here he is pictured at a WIVK event. (Photograph courtesy WIVK.)

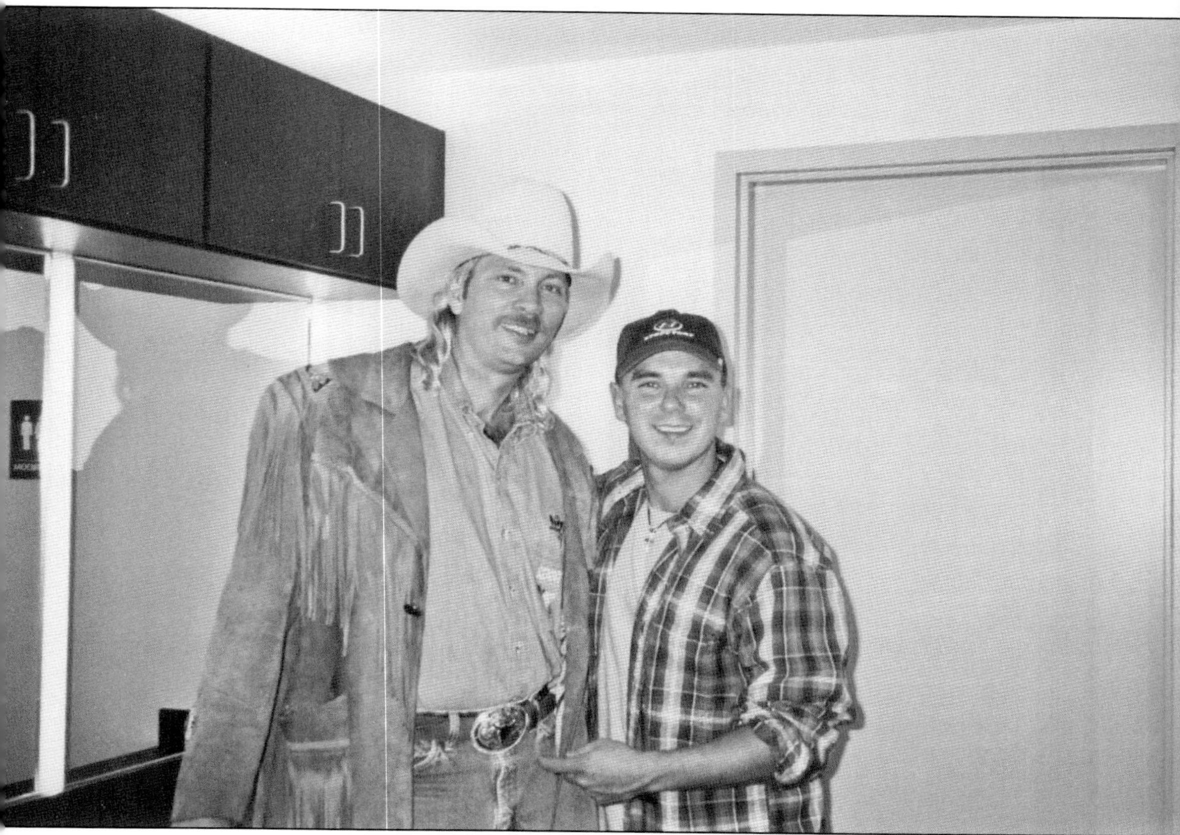

Because of his local roots in the city, Chesney (right) is considered a big part of the WIVK family as well. Here he is pictured with afternoon announcer Ted "Gunner" Ousley, whose afternoon drive program is the most popular in East Tennessee radio. His "Alan Jackson" looks have also landed him in a video or two. (Photograph courtesy WIVK.)

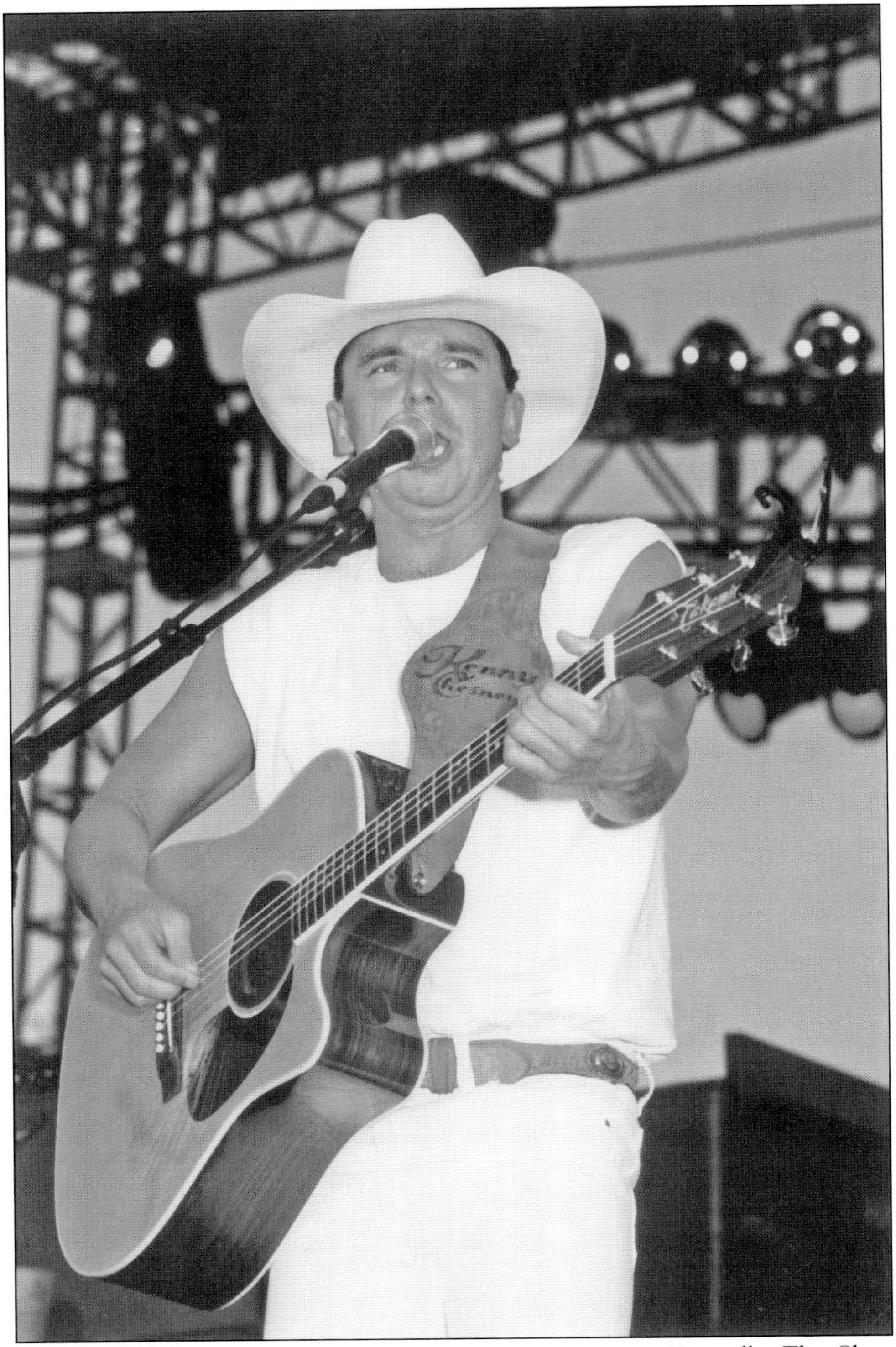

Here Kenny Chesney performs at a WIVK-promoted concert in Knoxville. The Chesney family still resides in Knox County, and Kenny is a frequent visitor when he to his home city. (Photograph courtesy WIVK.)

Five

Always Give Something Back

Jim Dick's "give something back" philosophy is one hammered home at the station through its involvement and leadership in countless charities and causes. There is not an announcer on the air or a newsperson that has not taken an afternoon, evening, or morning to talk to schoolchildren, church groups, and assemblies throughout East Tennessee. In some cases, this means lending a hand on backbreaking work to help out a cause or a listener. Many participate in charity basketball and softball teams against other media organizations in the region to help raise money. The employees of WIVK have always been its best representatives. Jim Dick always followed the rule that working in the media brings with it responsibilities to community and self that must be followed if one is going to be successful.

The station itself has done enormous work for the East Tennessee region and Knoxville through its legendary telethons, which have raised untold millions of dollars over its 55-year history for groups from St. Jude's to the Salvation Army and military family support groups.

Giving back to the community means more than raising money. WIVK has brought listeners some of the best free country music entertainment that can be found. The station sponsors numerous outdoor festivals or helps sponsor them by bringing in talent to boost local events.

The charities and causes the station has aided over the years are countless. Some are perennials and others are listeners or people in need, but the station cannot take all the credit. East Tennesseans take care of themselves, but they also respond overwhelmingly when there are crises or tragedies that strike the nation, and WIVK usually takes point. Floods, hurricanes, tornadoes, earthquakes, and wars have brought out the best in East Tennesseans.

In 2005, a year after Ted "Gunner" Ousley and the author returned from Iraq, Gunner and his wife, along with Blount County's First United Methodist Church of Maryville and WIVK, began working to bring eight-year-old Gufran Alayass and her family to East Tennessee for treatment of a severe case of spina bifida. The little girl was paralyzed from the waist down, and doctors worked to repair the damage to her spine and close the opening. The community welcomed the little girl and her family with open arms, and she recovered remarkably after her surgery.

Another highlight of Gunner's afternoon show was a feature called "Voices from the Front," started shortly after the September 11th tragedy. Families would tell him where their sons and daughters were serving in the war effort, and once a week, he was on the telephone with public affairs officers around the world setting up a call-in segment so the families could talk to their loved ones. It was touch and go because, for security reasons, he could not call them. They had to call the station, and they did. There were some tender moments and some funny ones with the program, but the WIVK segment was one of the most popular with both listeners and military personnel alike. With time differences, it was a program that sometimes found itself on in morning or during the midday drive.

The other announcers and station staff are just as dedicated. The Jim Dick creed of giving something back is still alive and well at the station, and it is not all speeches, work, and logistics. It can also be great fun.

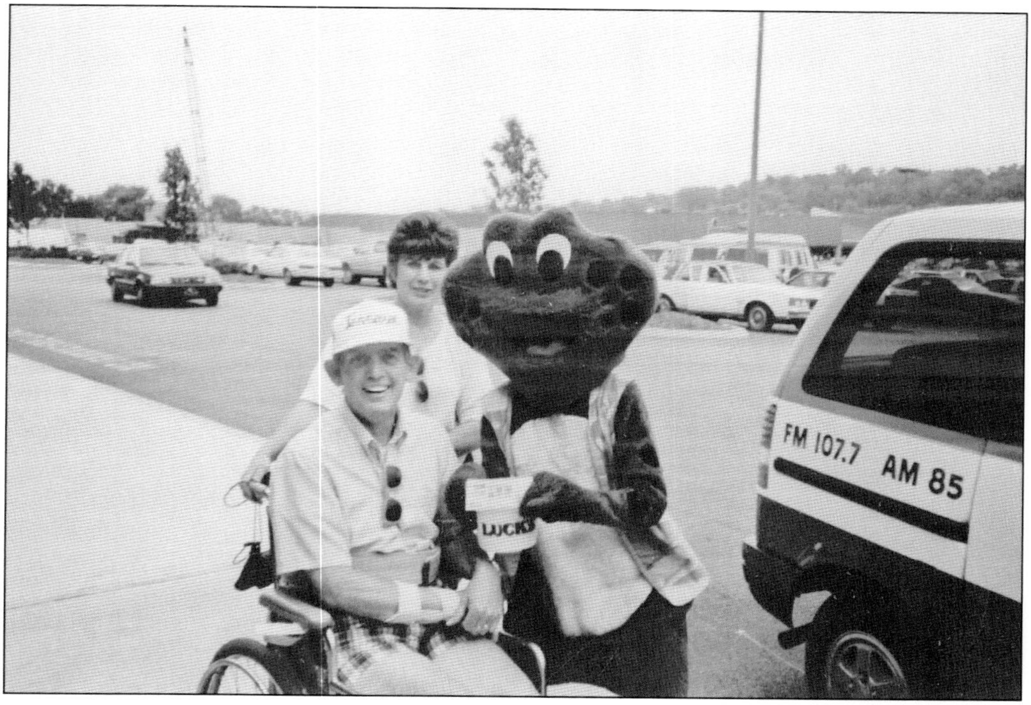

Wivik the Frog is the station's goodwill ambassador and the most recognized symbol of WIVK's community arm. The frog is always on the scene of station-sponsored events and covers for announcers when the schedule gets too busy. Below, the frog stands on the diving board doing a special on swimming safety for children. (Photographs courtesy WIVK.)

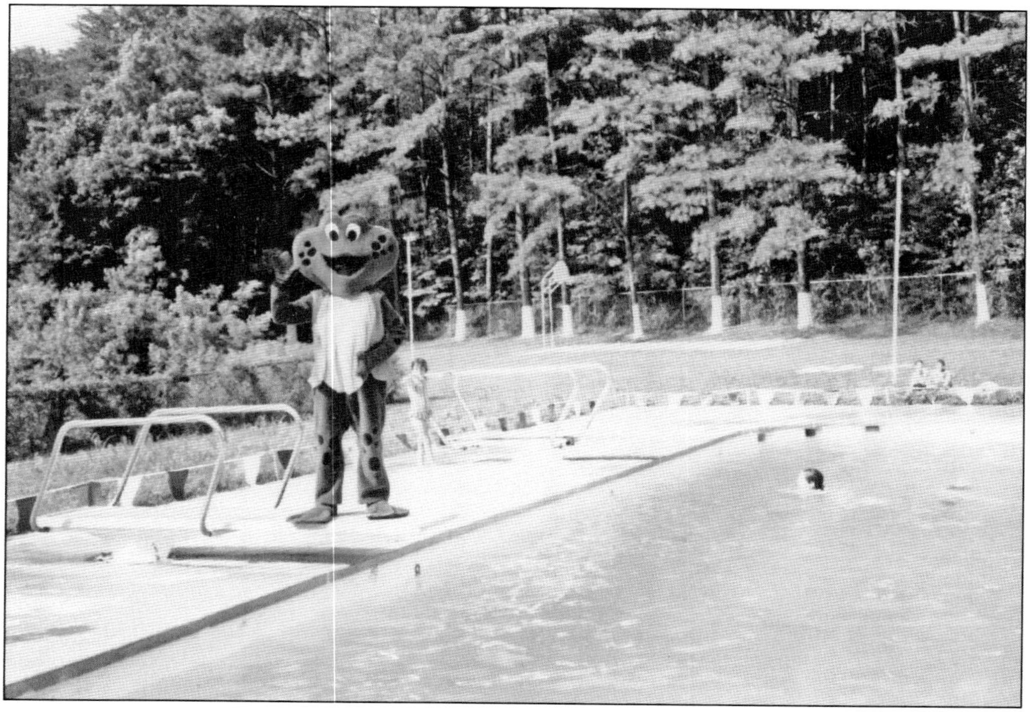

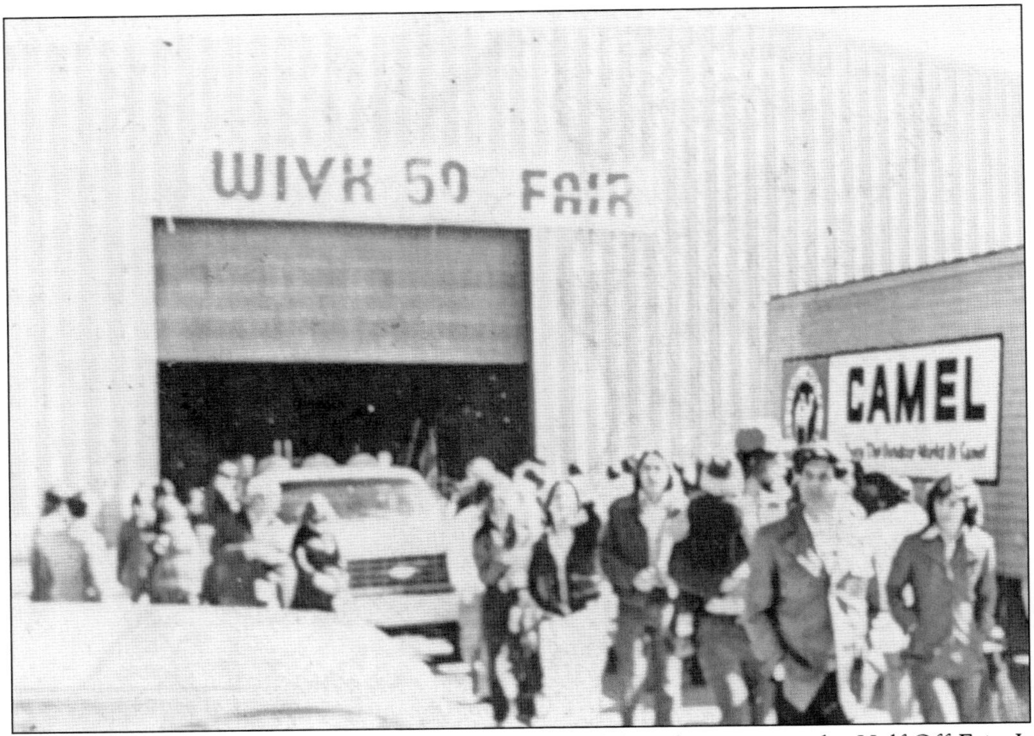

One of the most popular community events WIVK did in the past was the Half-Off Fair. It allowed listeners to purchase things from local retailers at 50 percent off retail cost. From clothes to sporting goods, merchants across East Tennessee participated in the event. The photograph below shows the crowds that turned out for the event.

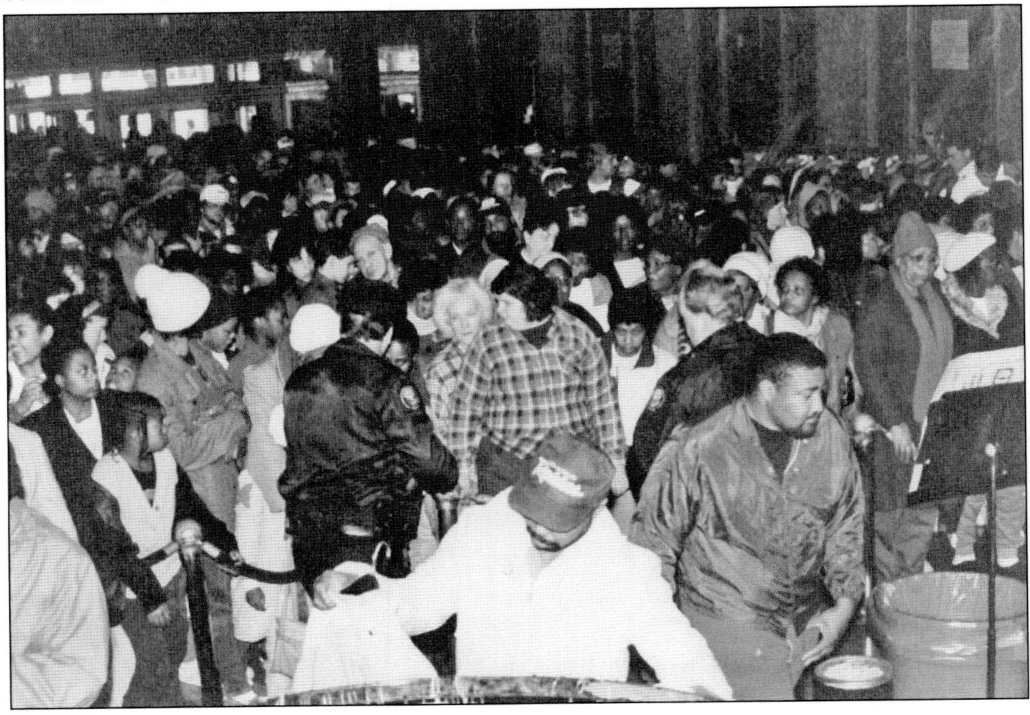

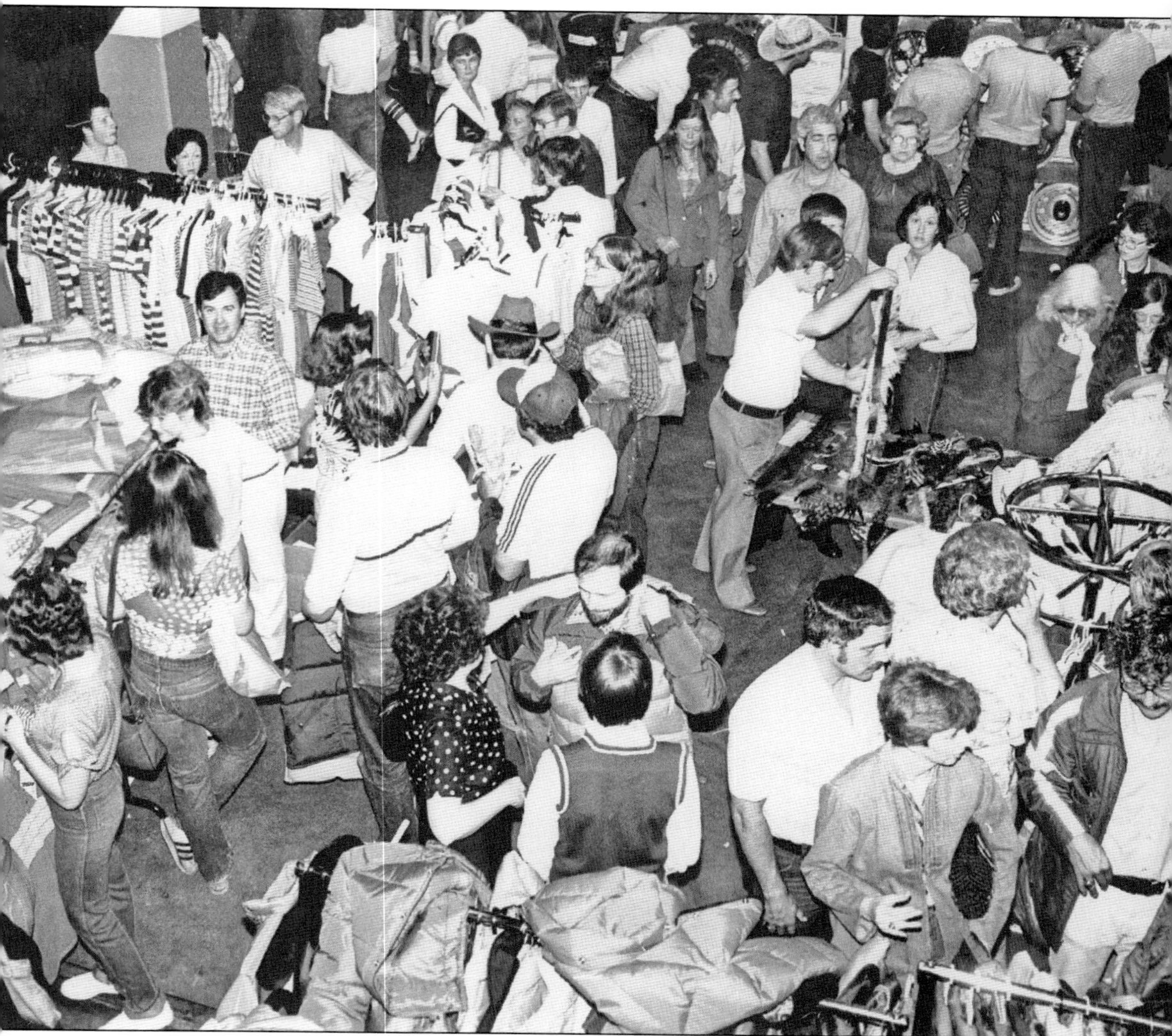

This is a photograph of one of the first WIVK Half-Off Fairs. The event was always promoted heavily on-air and drew people from across the region and sometimes the state looking for a bargain. They often found it at the Half-Off Fair, and the question was always, "When is the next one?"

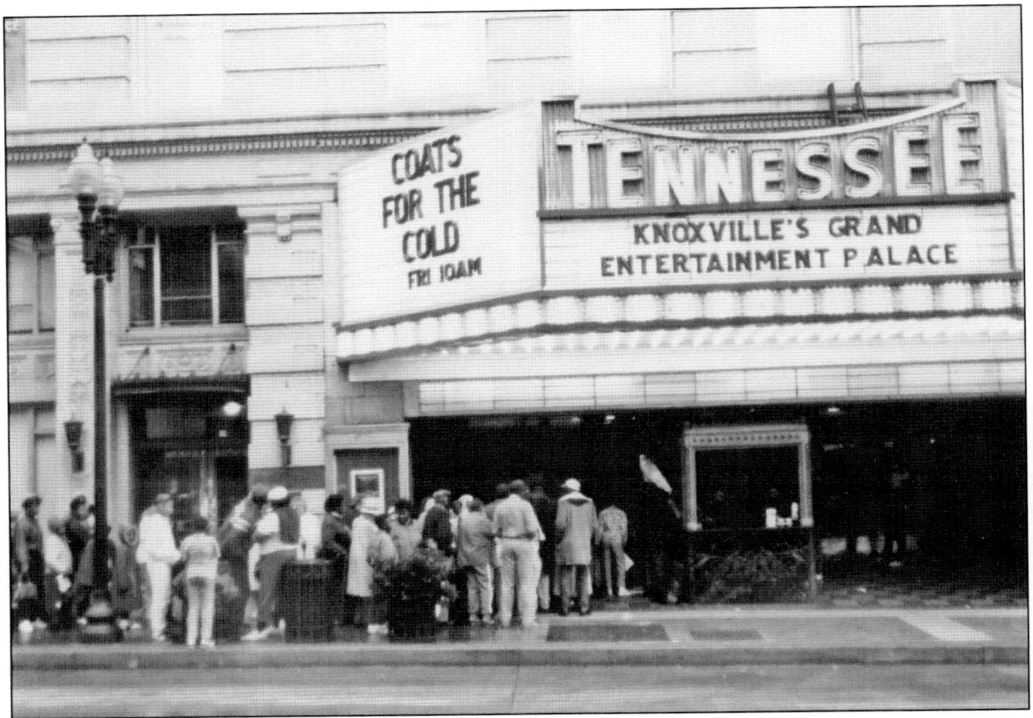

Afternoon-drive announcer Ed Brantley created one original charitable cause for WIVK called "Coats for the Cold." He was on his way into the station one day when he noticed a man in freezing weather who said he did not have a coat. Brantley came up with the idea the to do "Coats for the Cold" to help the less fortunate. Above, the line for those needing coats stands outside Knoxville's Tennessee Theater on Gay Street. Below is an image that reflects how long that line can be at times. (Photographs courtesy WIVK.)

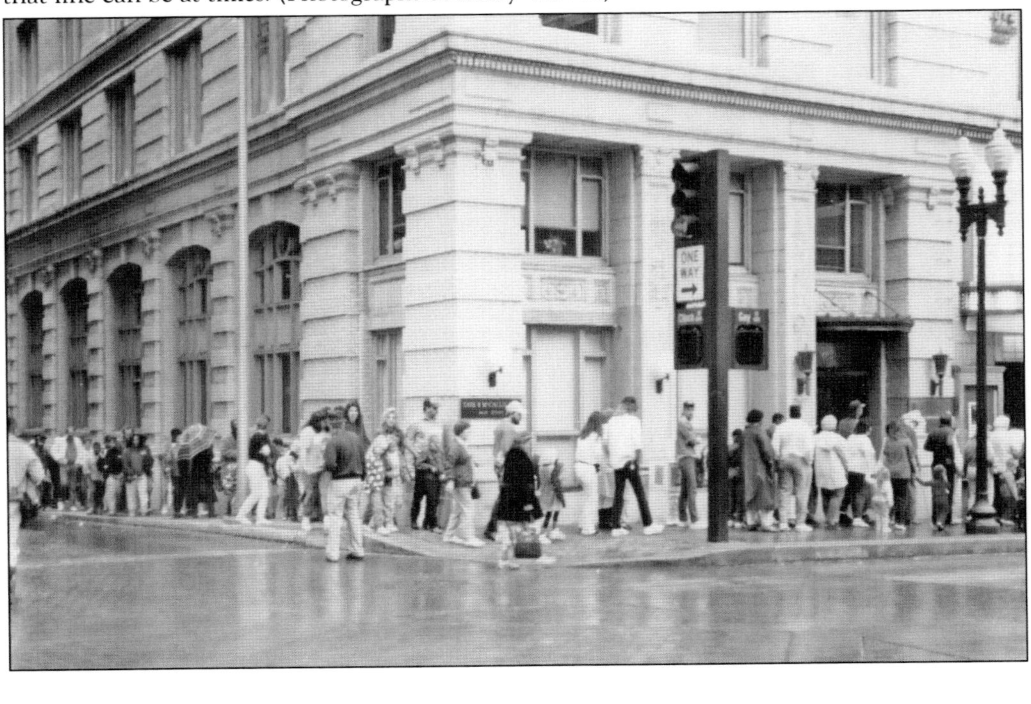

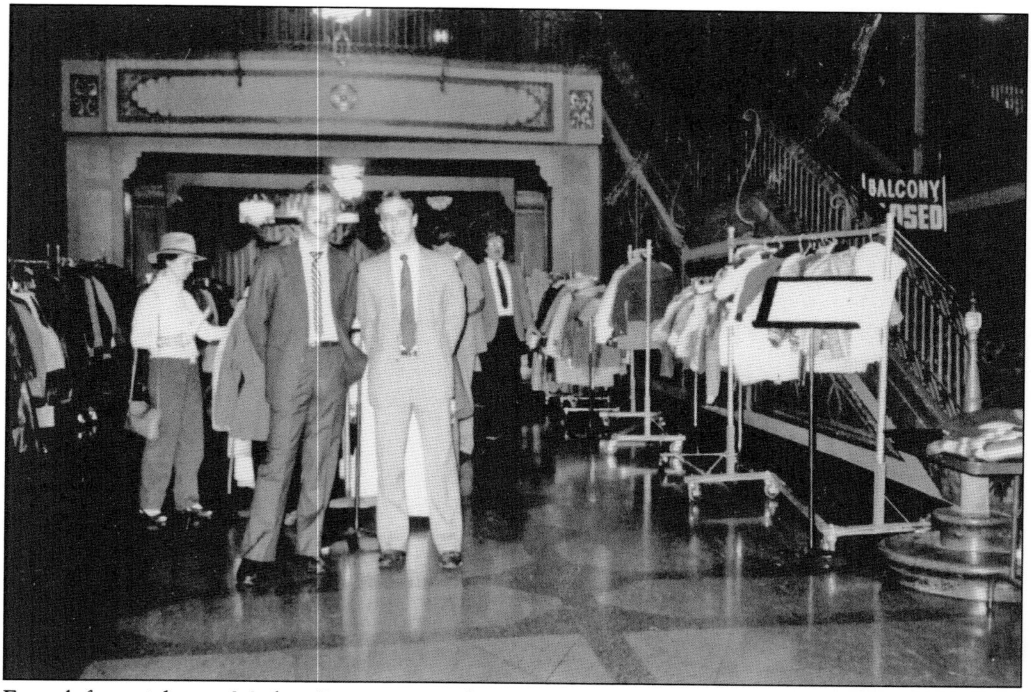

From left to right are Mickey Dearstone and "Coats for the Cold" creator Ed Brantley, who take time from the event to get their photograph taken. (Photograph courtesy WIVK.)

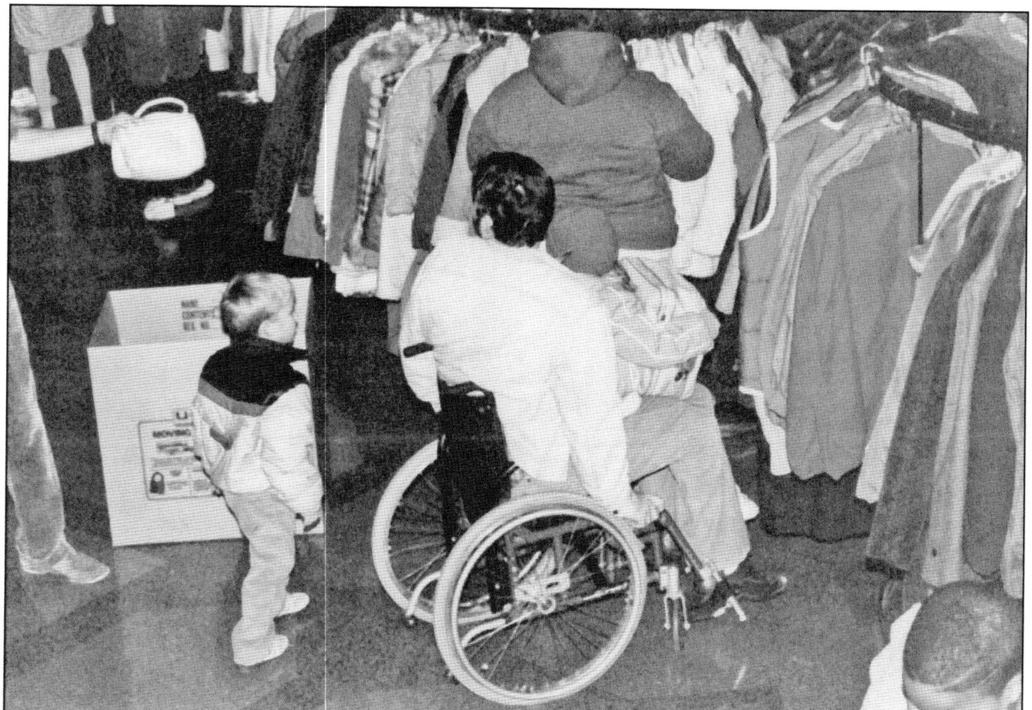

In some cases, entire families are in need of coats, as East Tennessee winters can be unpredictable and bitter. "Coats for the Cold" is still one of the station's annual charitable drives.

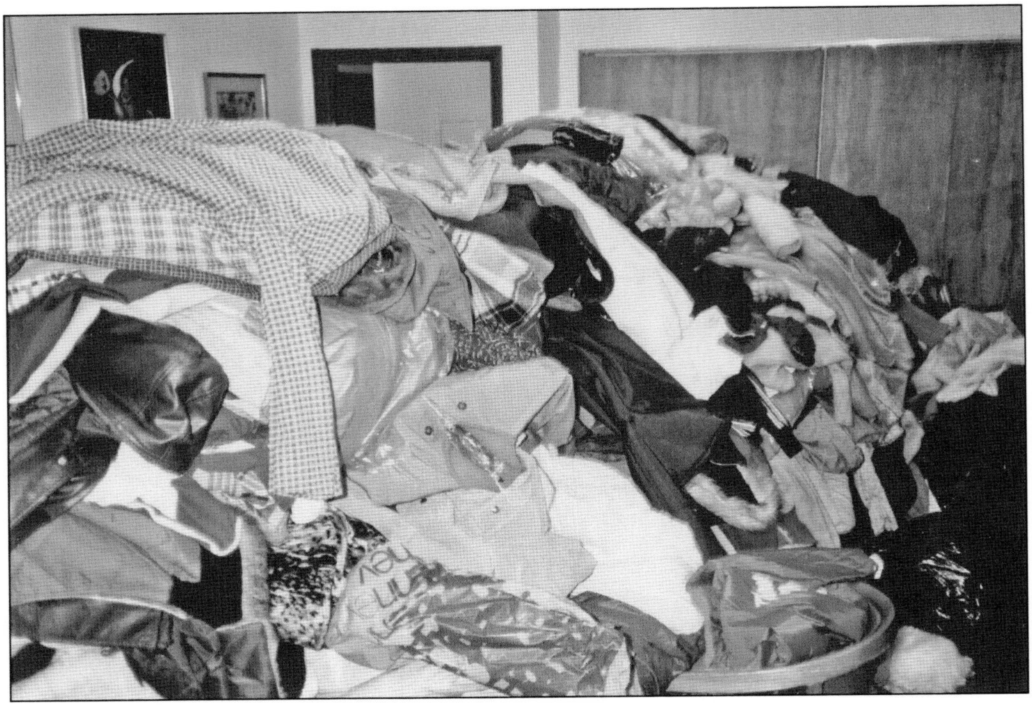

WIVK listeners donate coats for the annual drive, and station personnel are often overwhelmed by the number of coats given to the cause each year by East Tennesseans. Below is an image of the coats on racks waiting for those that need them. Those coats remaining after the "Coats for the Cold Day" are donated to local charities. (Photographs courtesy WIVK.)

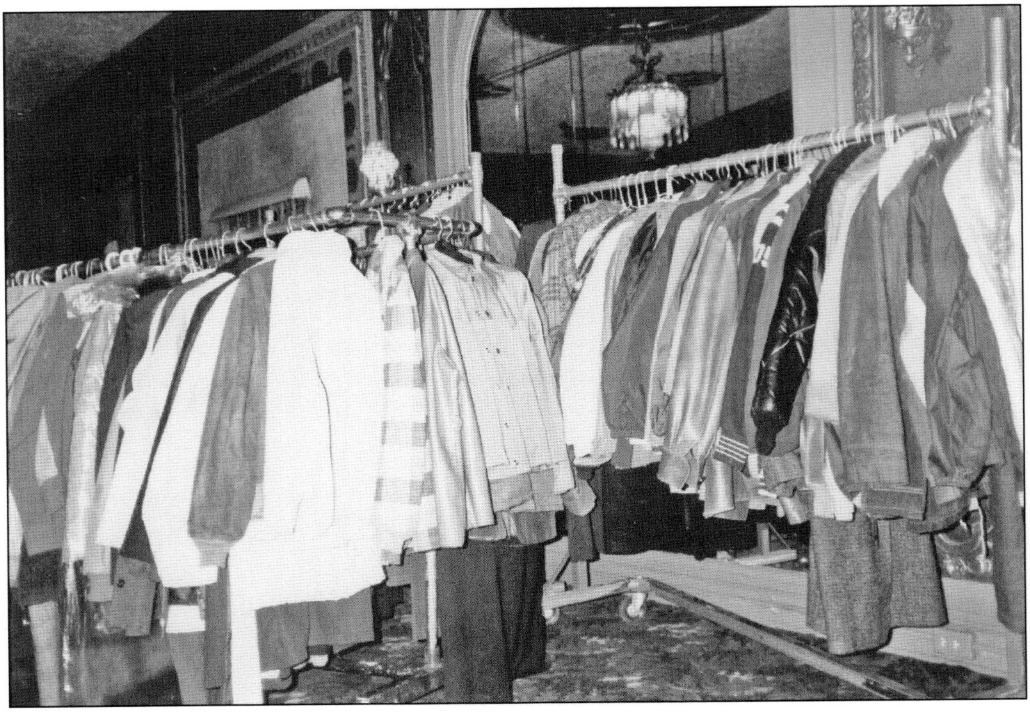

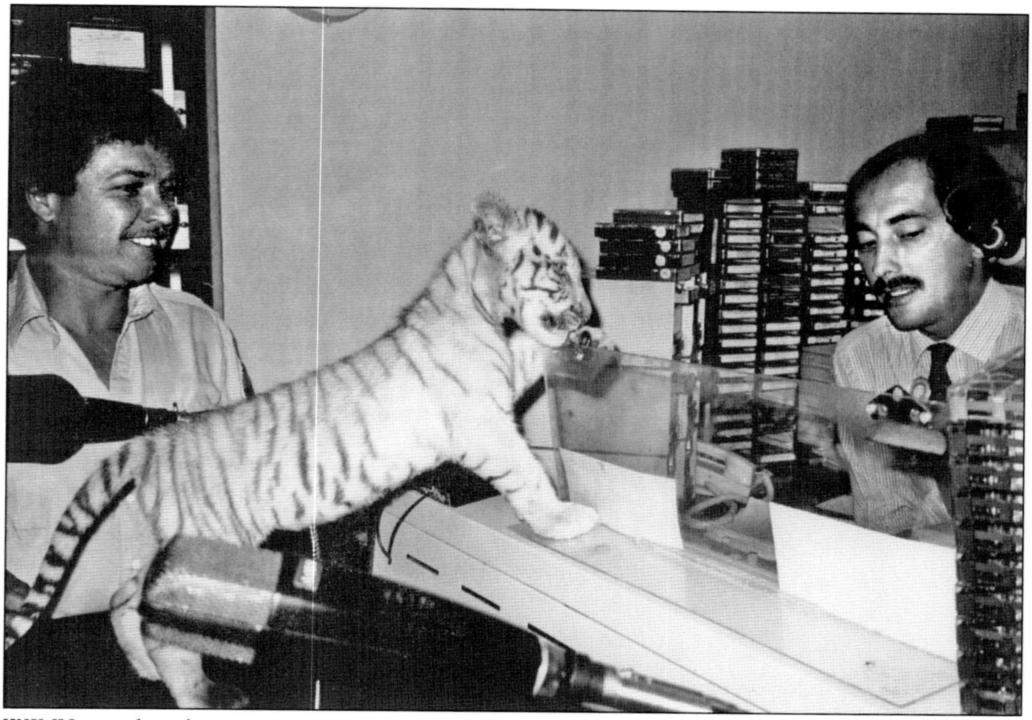

WIVK was also a big supporter of the Knoxville Zoo from its origins in the 1960s. Here announcer Bob Thomas (right) and a zoo animal handler do an on-air "Name the Tiger Cub" contest.

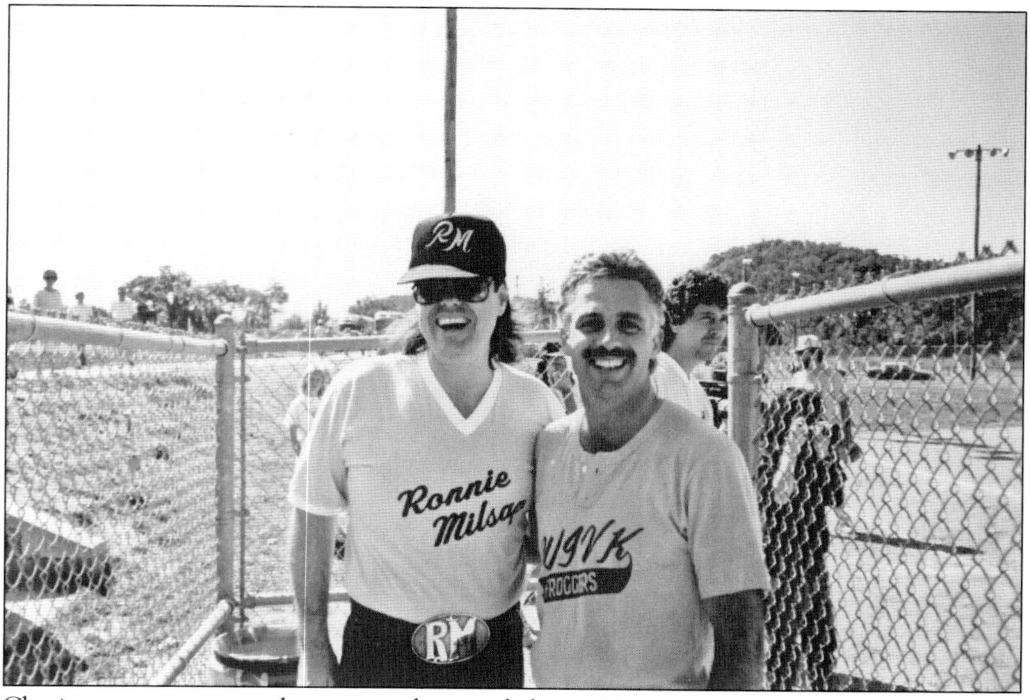

Charity sports teams are always a popular crowd pleaser. Here Ronnie Milsap (left) stands with a WIVK employee during a lighter moment in the competition. (Photograph courtesy WIVK.)

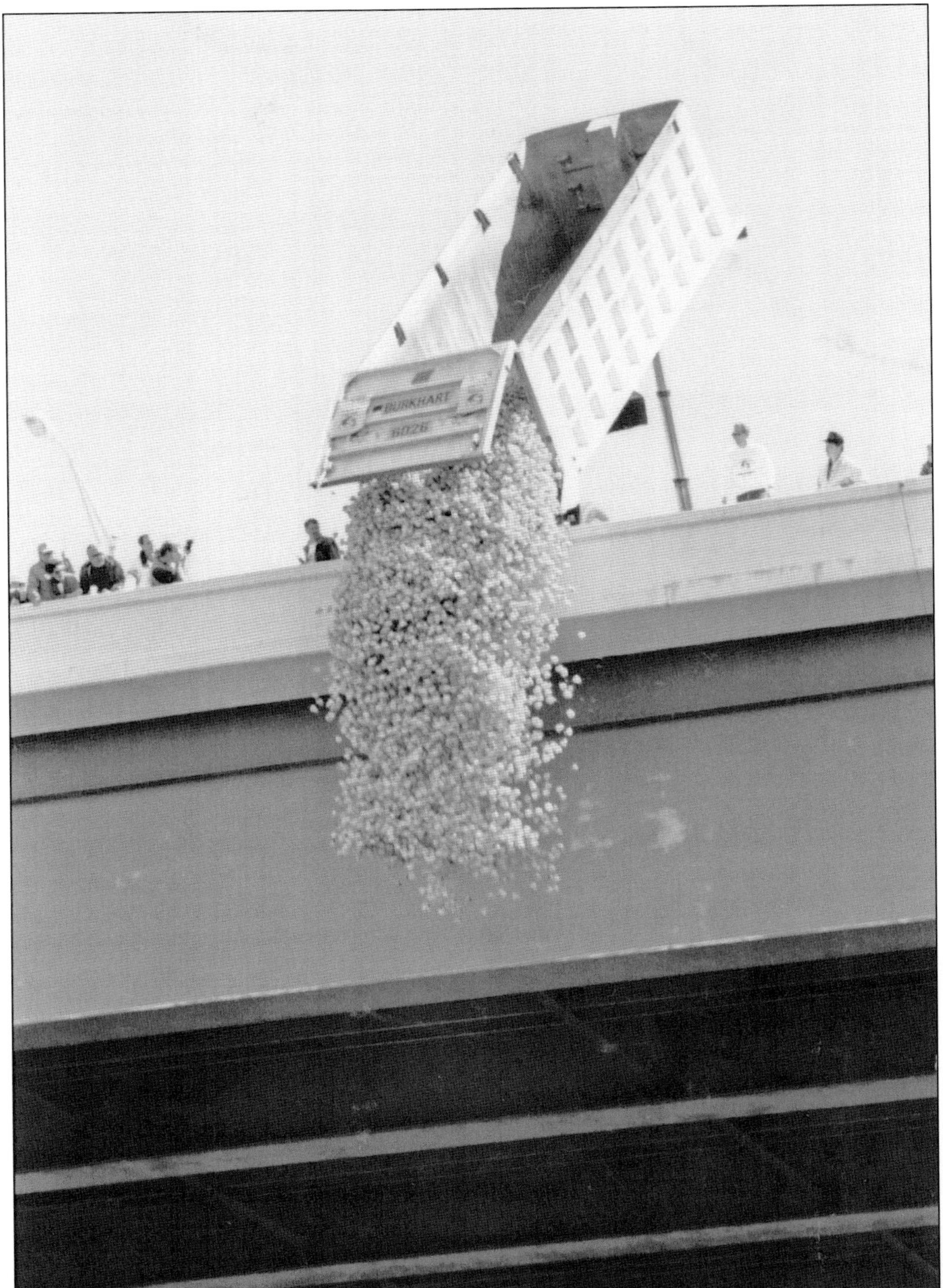

WIVK was also a big supporter of the annual Knoxville Rubber Duck Race. The ducks are numbered and sold for $5 each and then dumped off of the Gay Street Bridge. The person who has the winning duck across the finish line gets a huge prize. All the money raised from the event benefits the Boys and Girls Club of the Tennessee Valley. (Photograph courtesy WIVK.)

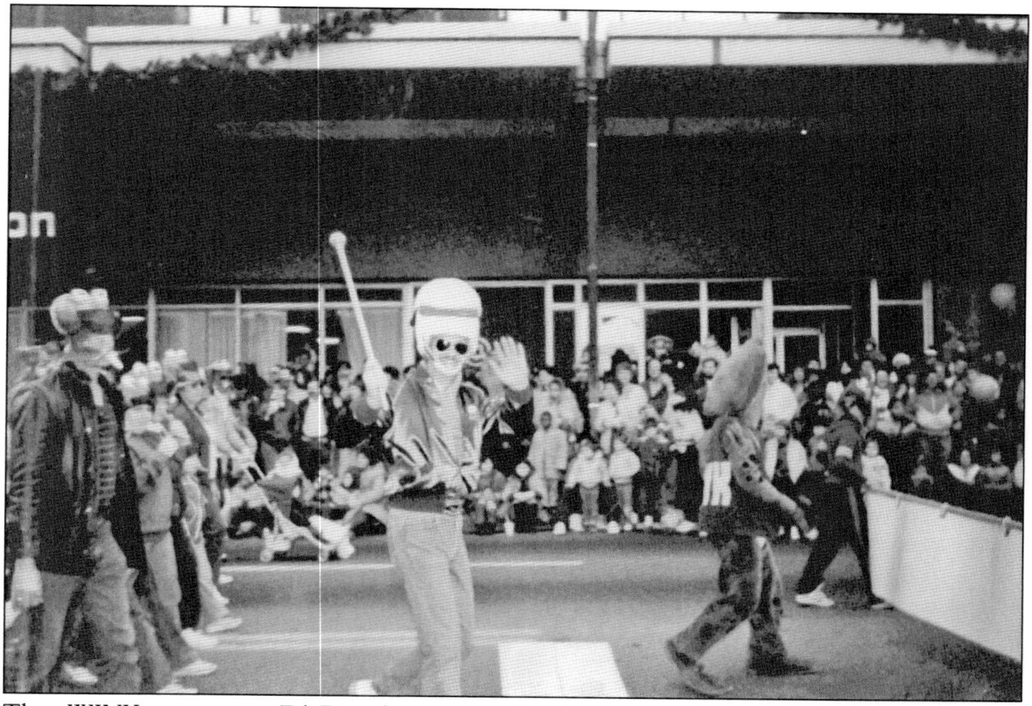

Then-WIVK announcer Ed Brantley acts as the drum major of the 1986 WIVK all-kazoo marching band, leading them down Knoxville's Gay Street at the annual WIVK Christmas parade. The annual parade is a city-wide event that draws people from across East Tennessee. Pictured below is the signage and Wivick the Frog helping lead the band. (Photographs courtesy Ed Brantley.)

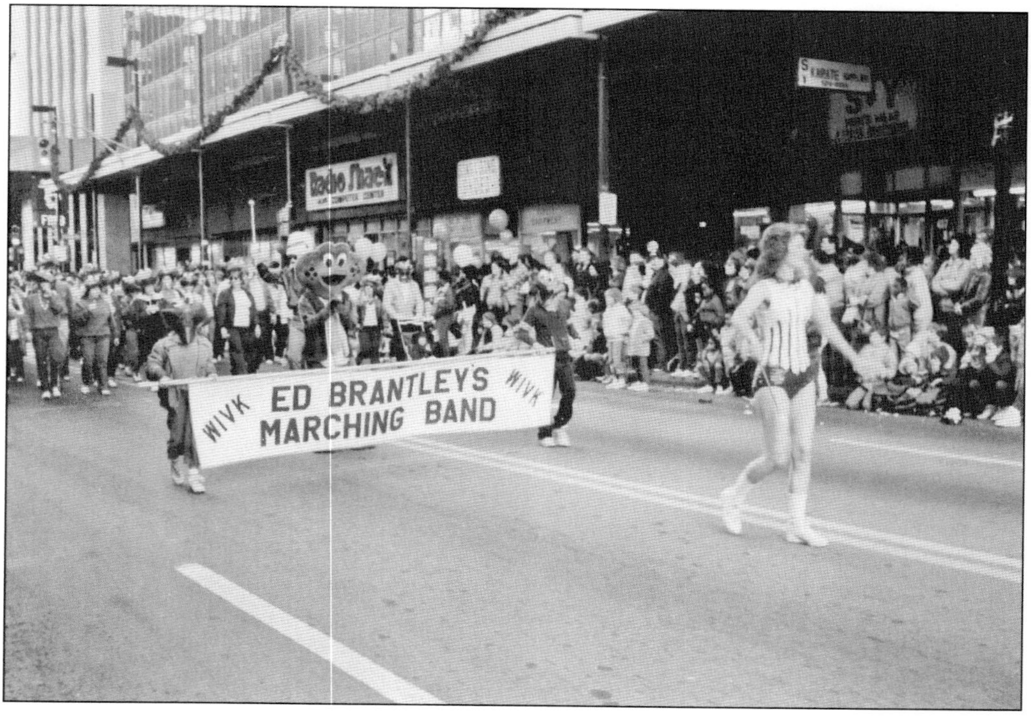

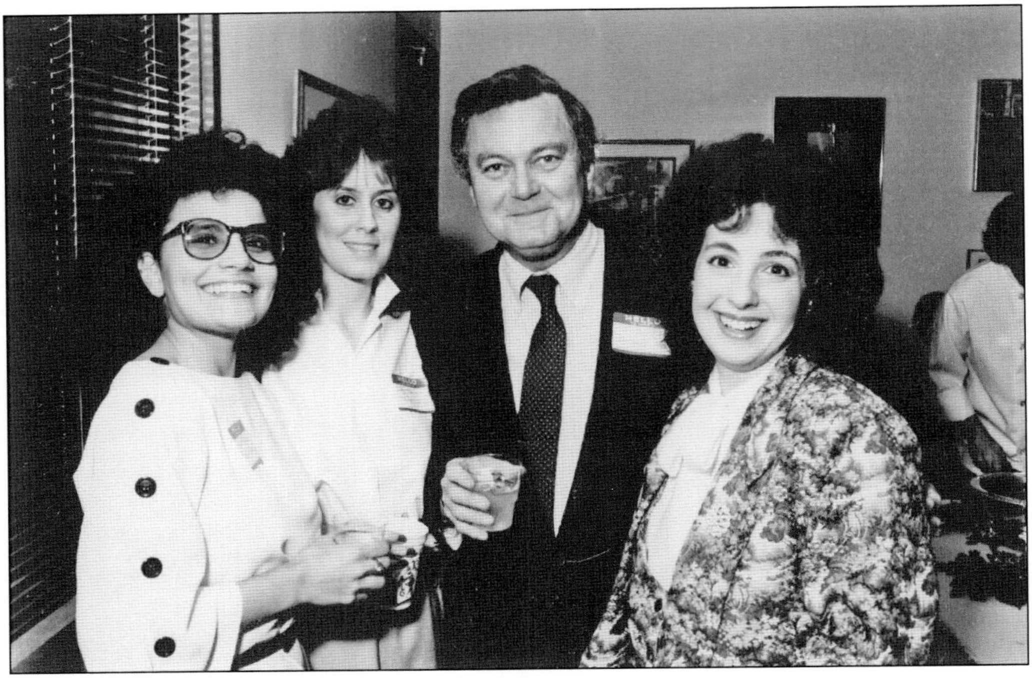

Giving back to the community also includes the popular client parties hosted by WIVK. Third from left is legendary Knoxville broadcaster and advertising salesman Jim Early. He would later go on to found Earl and Associates Advertising. Below are WIVK announcer Bob Thomas and his wife, Kim, at a WIVK function. (Photographs courtesy WIVK.)

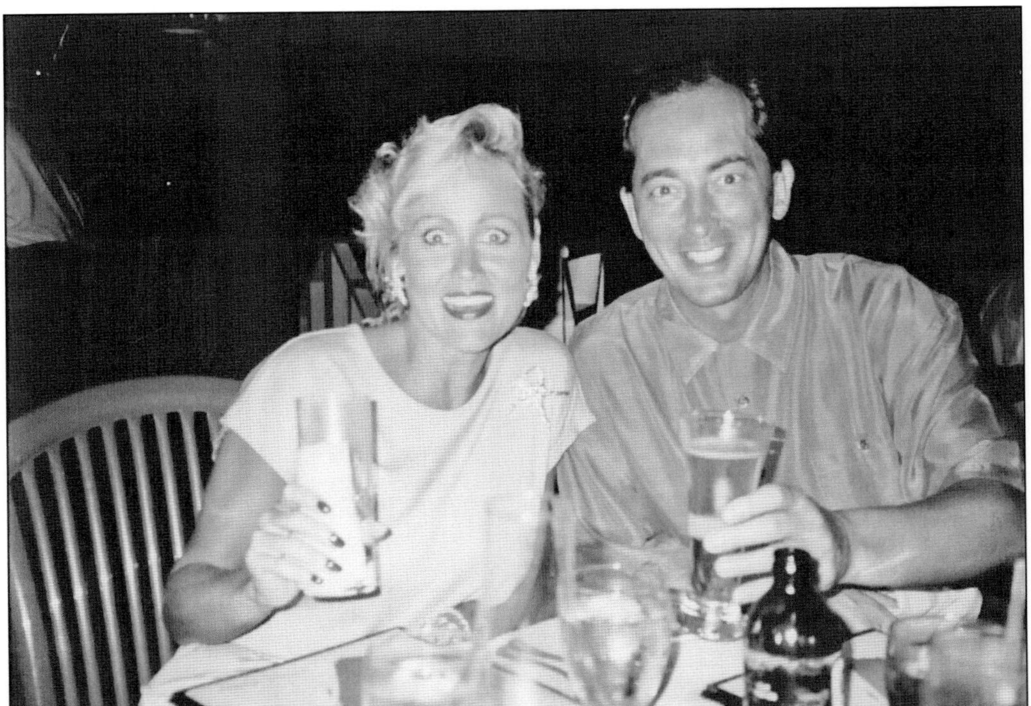

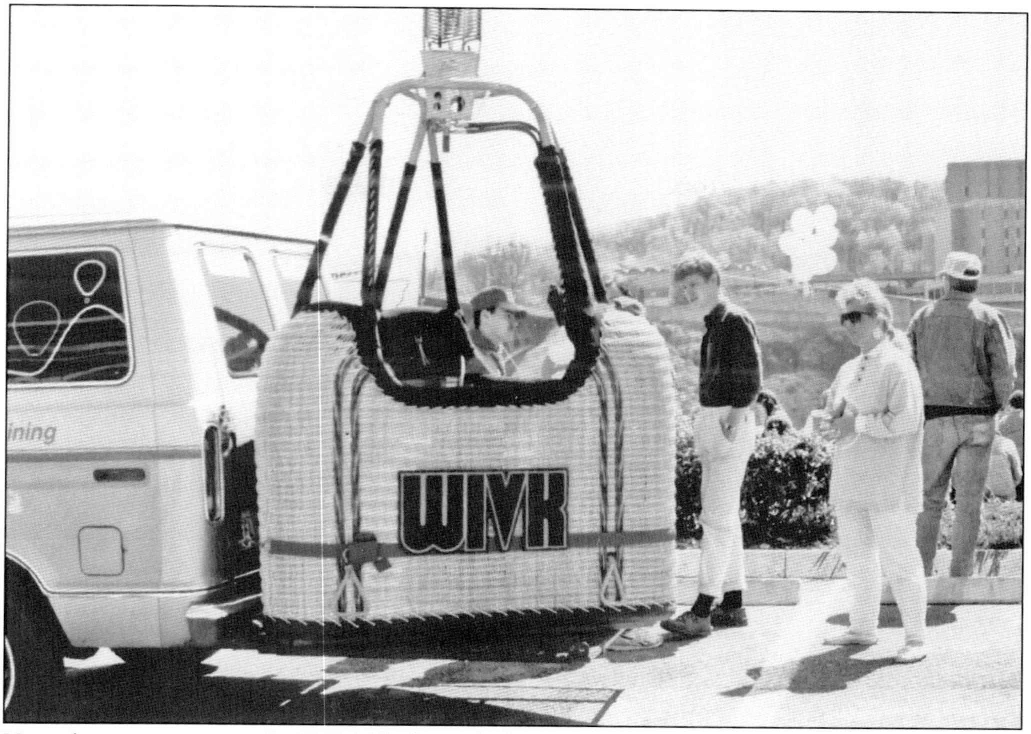

Here the crew prepares the WIVK balloon for lift-off. The balloon was a popular attraction for listeners. (Photograph courtesy WIVK.)

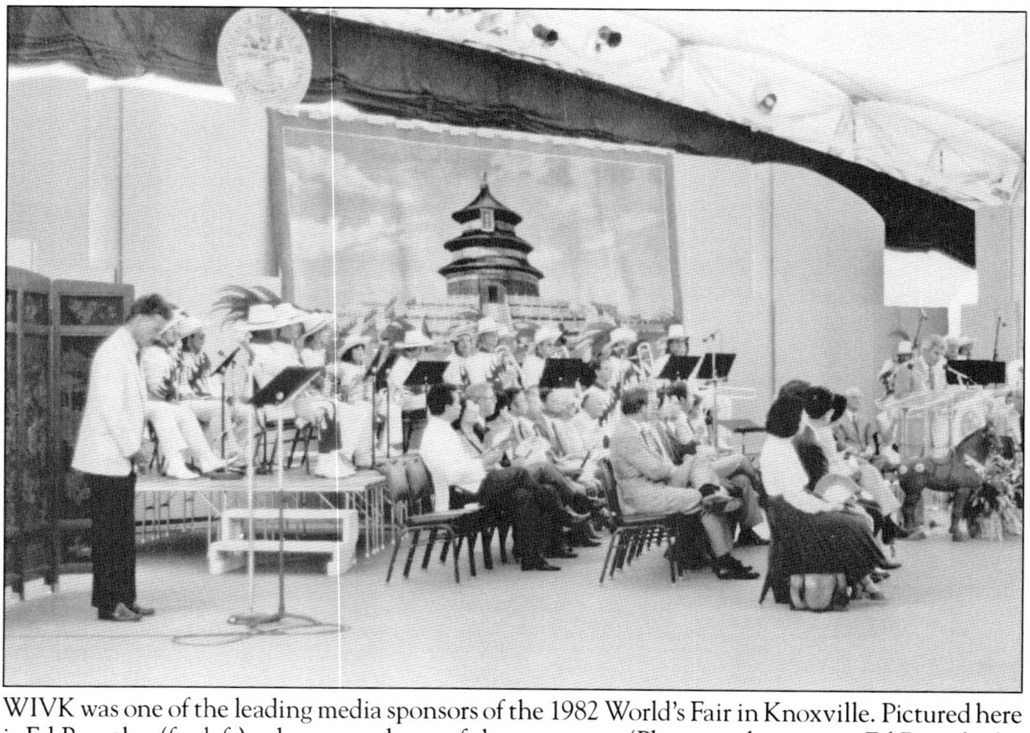

WIVK was one of the leading media sponsors of the 1982 World's Fair in Knoxville. Pictured here is Ed Brantley (far left), who emceed one of the programs. (Photograph courtesy Ed Brantley.)

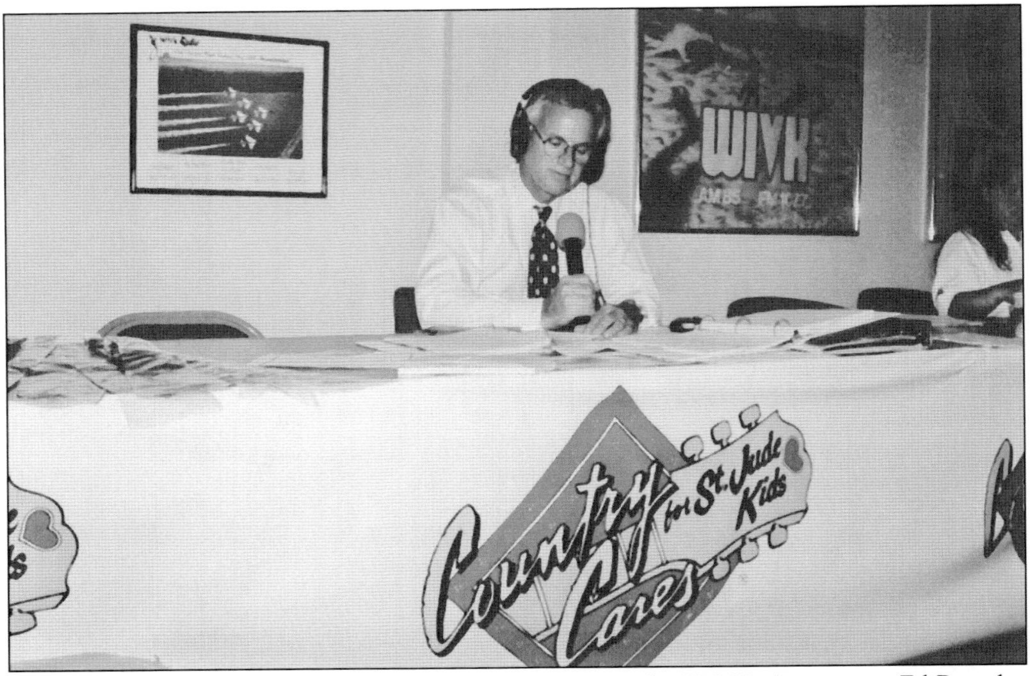

Helping St. Jude's Hospital is also a big charity promotion for WIVK. Announcer Ed Brantley does his afternoon show from the "Country Cares for St. Jude Kids." The telethon raised thousands of dollars for the hospital. (Photograph courtesy Ed Brantley.)

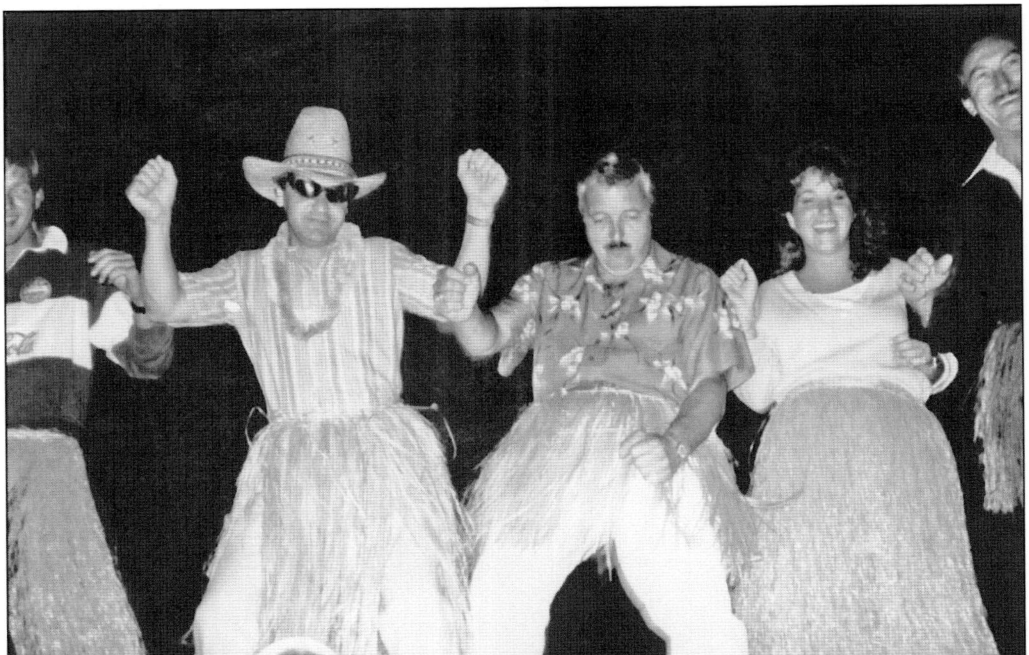

WIVK general manager Bobby Denton (center) is seen at Knoxville's Riverfeast Show trying to show he can hula for the crowd.

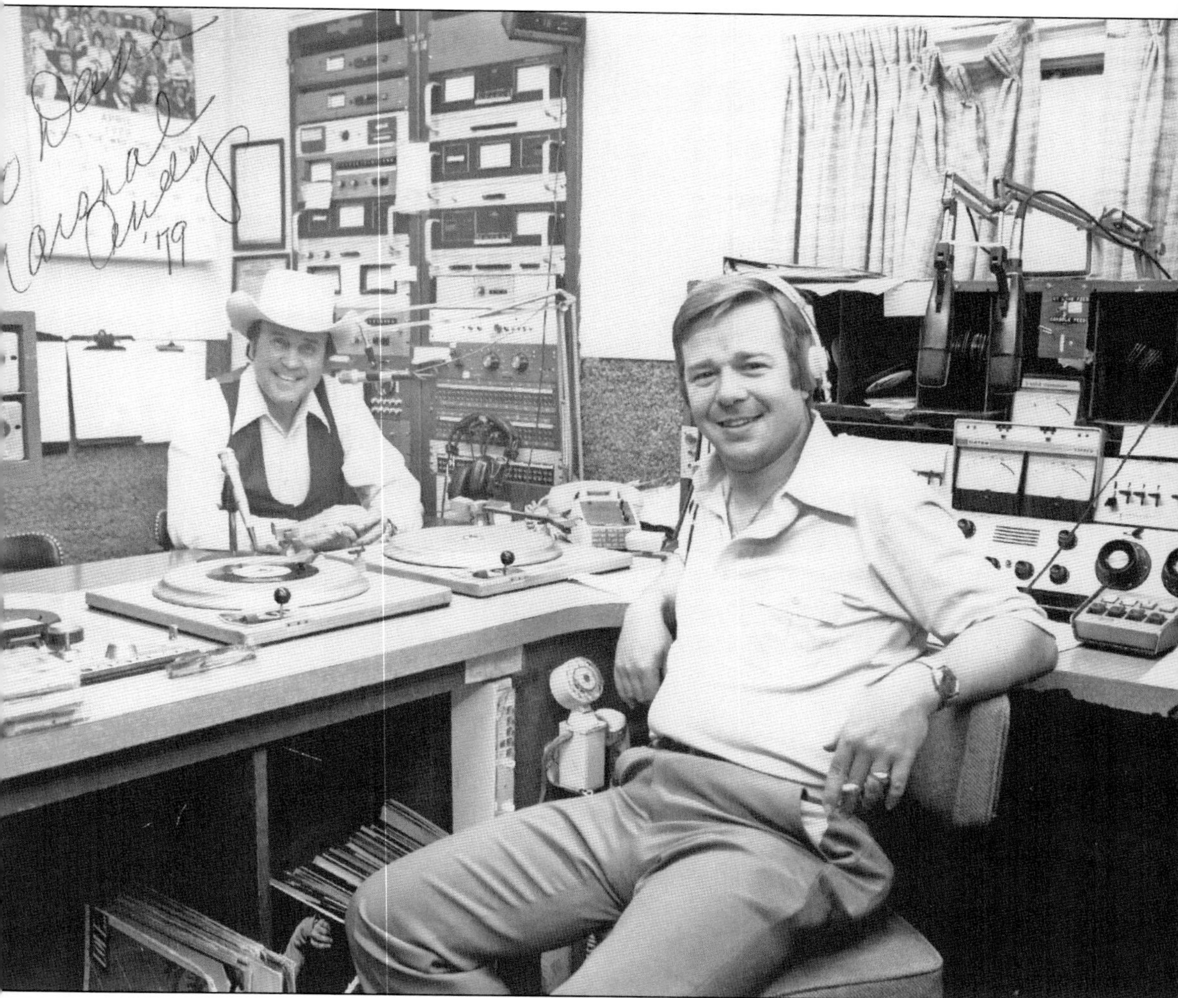

Here Dave Young (right) and Andy Smalls, known as "Marshal Andy" on Public Television of East Tennessee, pose for a photograph in the WIVK control room. The station has always been a big supporter of public television in the region.

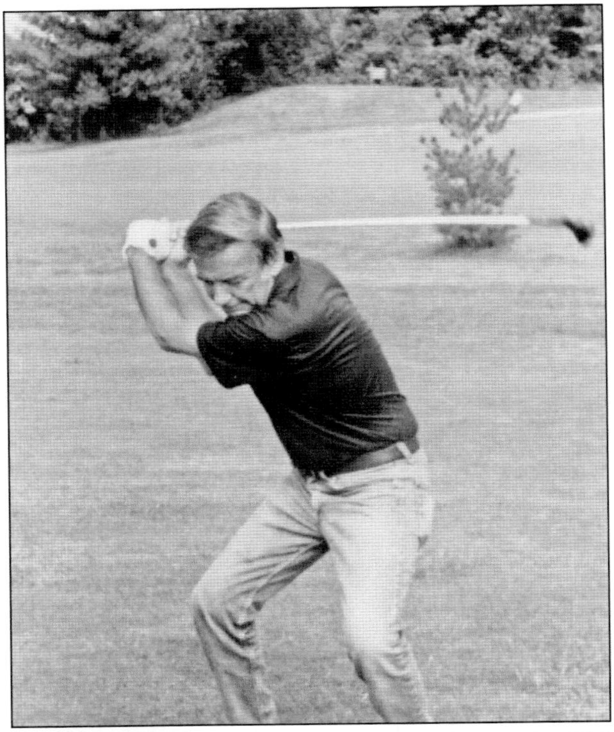

Claude Tomlinson takes a swing on the course. One of the leading places where many charity fund-raisers are held is on local golf courses, and WIVK always fielded a competitive team to help the cause. Below, the team is wearing the company colors. From left to right are Jim Christensen, David Henderlight, Bobby Denton, Mickey Dearstone, and Frank Nystrom.

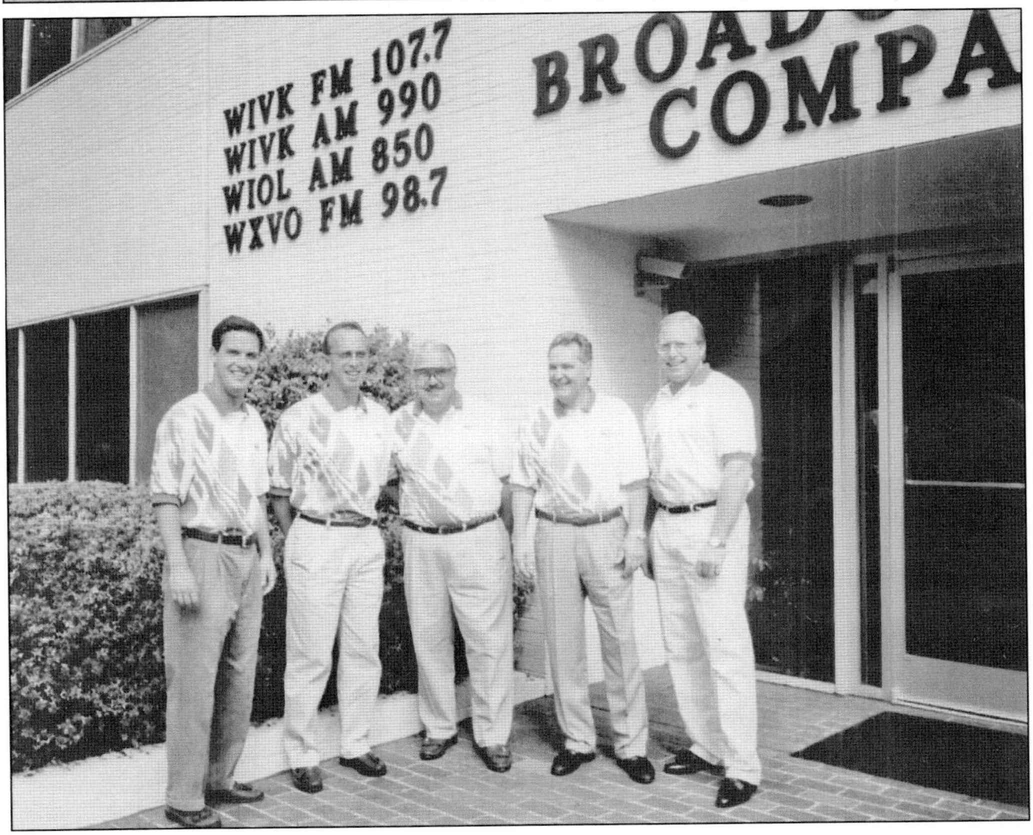

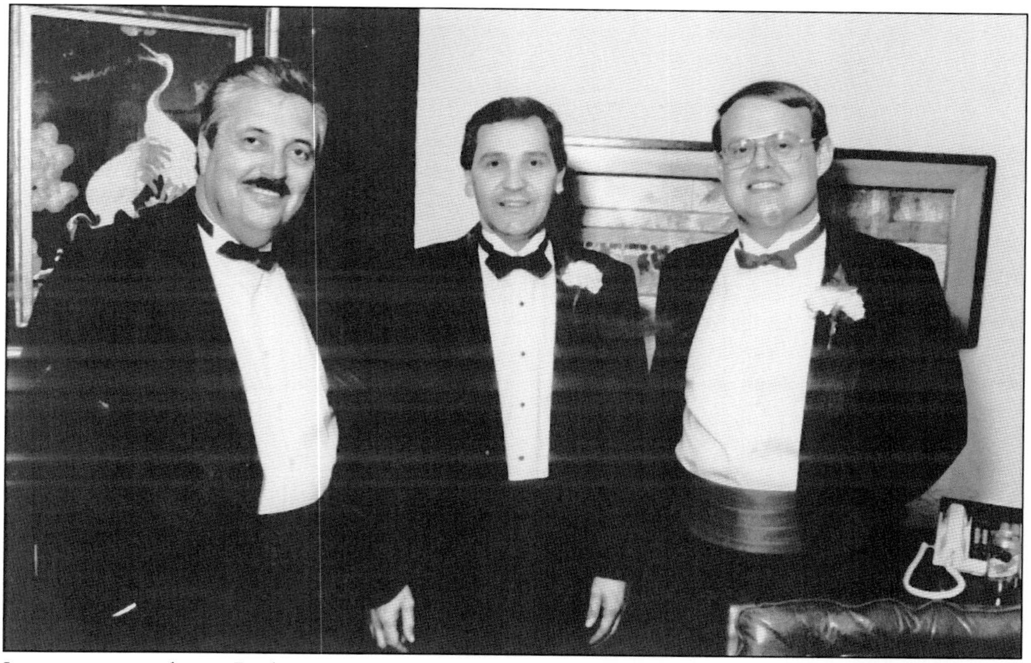

Image is everything. Radio station employees are usually known for casual attire, but WIVK has hosted a number of black-tie events through the years. From left to right are WIVK general manager Bobby Denton, David Henderlight, and current operations manager Mike Hammond. (Photograph courtesy WIVK.)

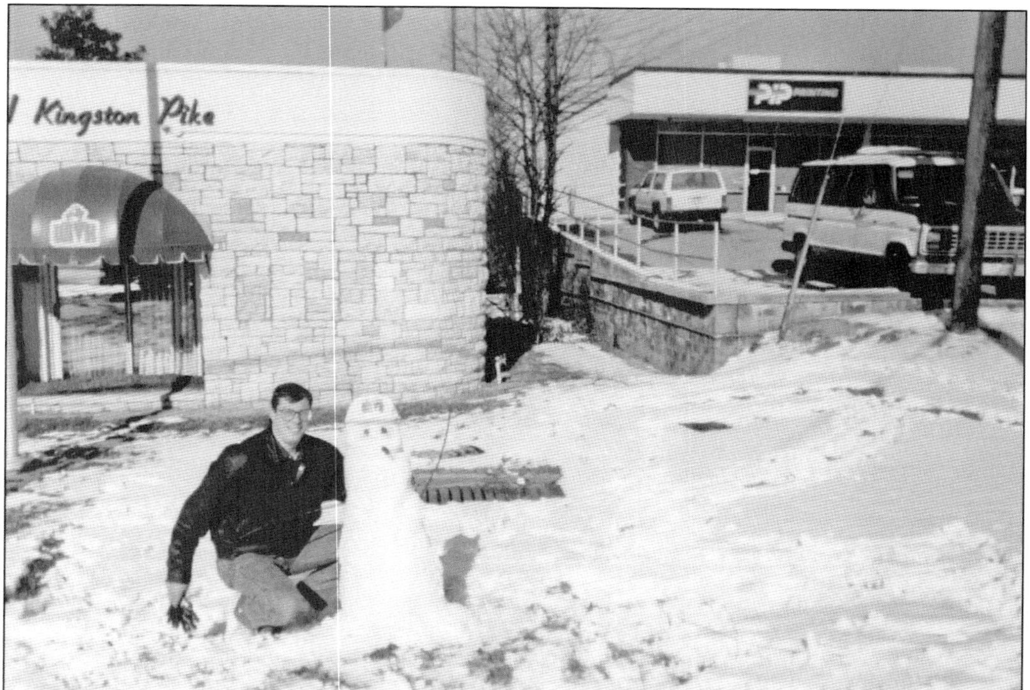

Tim Wall takes time out to build a snowman in front of the WIVK studio at 6711 Kingston Pike for listeners of the *Wilhite and Wall Show*.

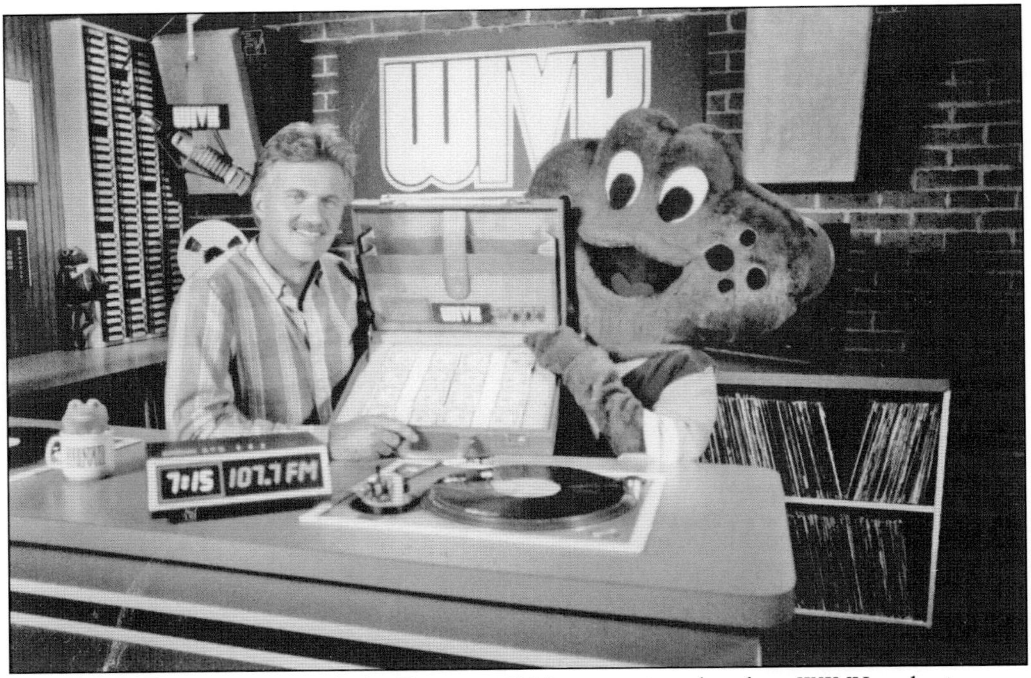

This is Ed Brantley and Wivick the Frog in a 1980s promotion shot for a WIVK cash giveaway that was popular among station listeners. (Photograph courtesy Ed Brantley.)

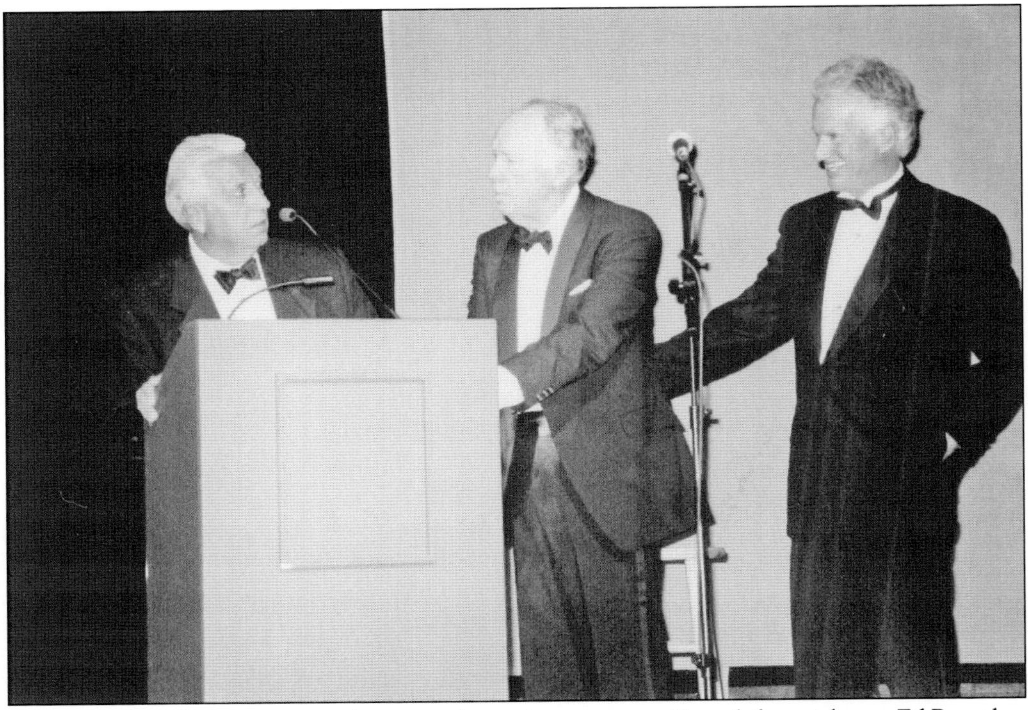

Giving back to the community also means honoring your own. From left to right are Ed Brantley, Jim Dick, and Bobby Denton at a gathering honoring the station's founder. (Photograph courtesy WIVK.)

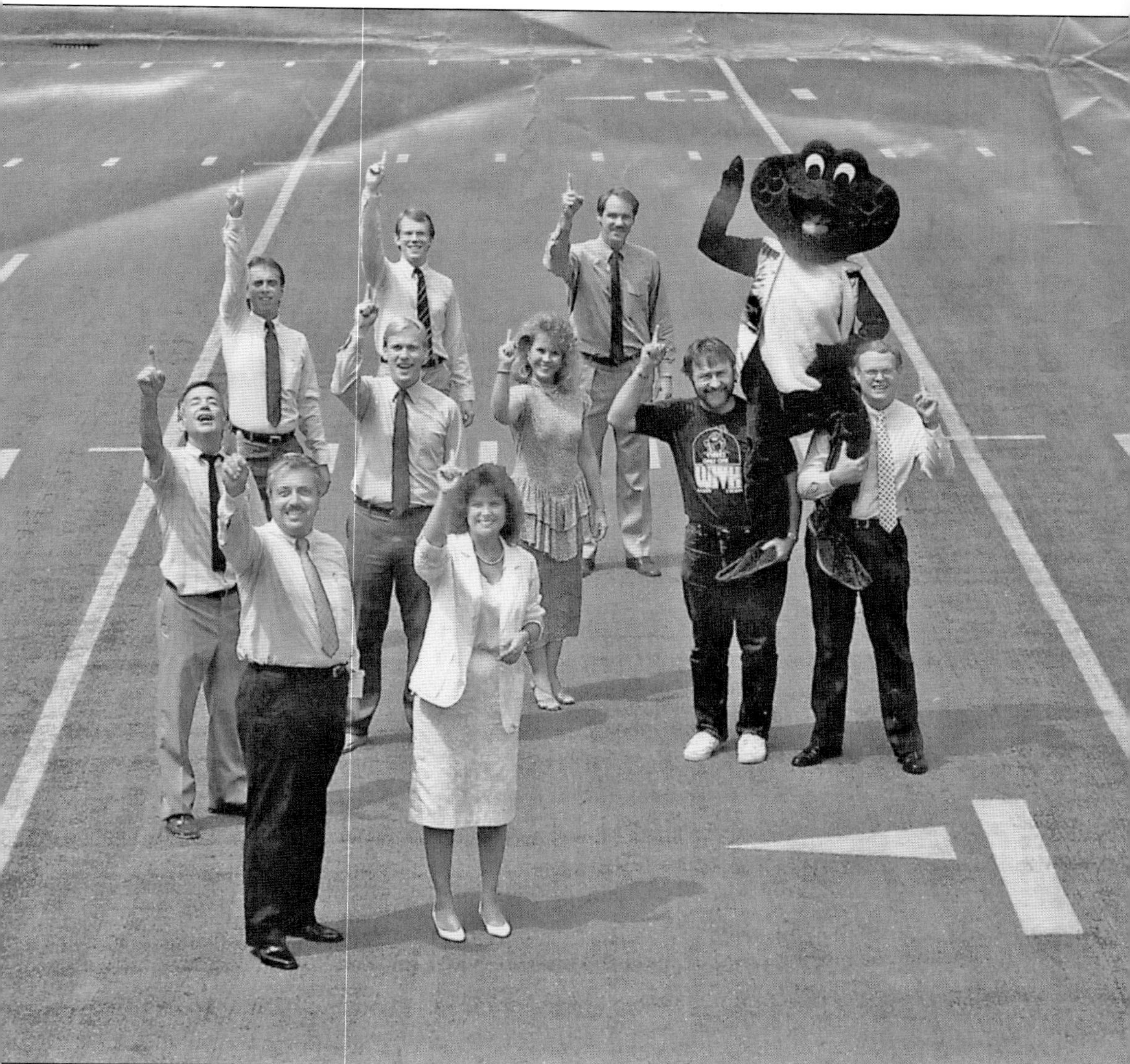

While University of Tennessee sports are a big ratings-getter for the station, it can't be understated that WIVK has been a huge contributor to the university's academic programs. This is a group shot taken on Neyland Stadium's Shield–Watkins Field when WIVK was named America's most listened to radio station. From left to right are (first row) Bobby Denton and Lisa Patton; (second row) Dave Young, Jim Donovan, Laura Paul, Dan Bell, Wivick the Frog, and Mike Hammond; (third row) Mickey Dearstone, Beecher Bailey, and Steve Armistead.

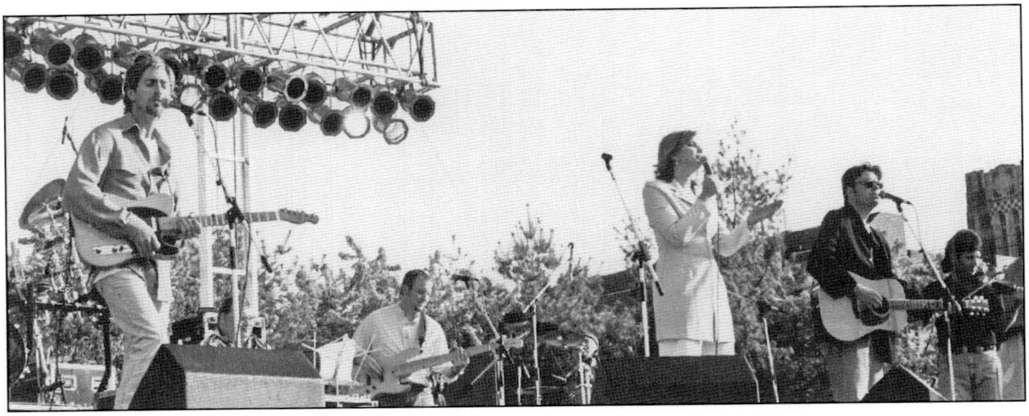

The Foothills Fall Festival is a three-day festival held in nearby Maryville, Tennessee. The annual event was noted by *USA Today* as one of the 10 best fall festivals in the nation. WIVK is one of the sponsors of the event, which started in 1999. Seen below is the WIVK stage, where thousands of listeners and country music fans turn out to hear the latest bands. (Photographs courtesy WIVK.)

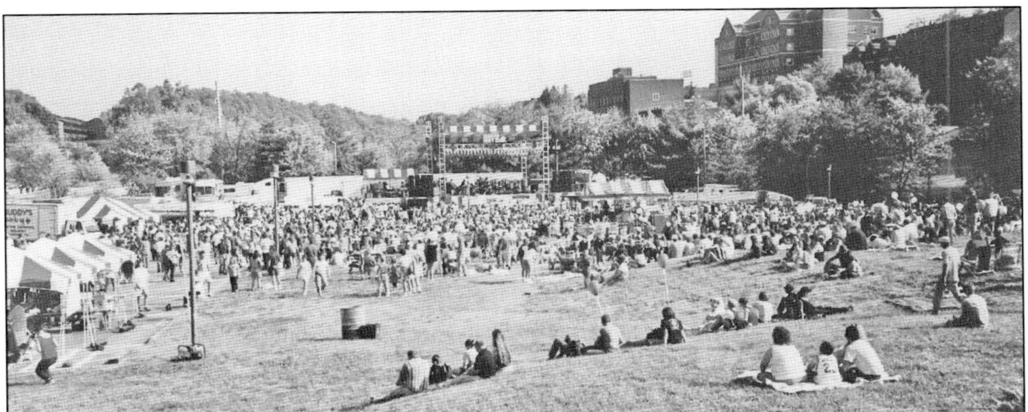

Six

WIVK FROM THEN TO NOW

From 1953 to 2008, WIVK has gone from a small studio on Gay Street to a first-class radio station. There have been some stumbles along the way, but none the station could not recover from. Seven times it has been named the Country Music Radio Station of the Year by the Country Music Association; the Academy of Country Music has named it Station of the Year five times; and the station has won the coveted Marconi Award four times. Add 8 awards from the Country Radio Broadcasters, 9 Edward R. Murrow Awards, 20 Associated Press awards, and countless honors from the Society of Professional Journalists, and WIVK is a full-service radio station that is like no other in the United States.

It is a demand for excellence that few companies, regardless of the industries they are a part of, can match. It did not come easy. The station grew in spurts and followed the old adage of never paying for the same real estate twice. The real credit of a radio station's success, however, is the advertising sales representative listeners never see. Selling air time on a clock evolved from the early days of block programming to now, where every second of every broadcast day is measured and someone somewhere knows who and how many are listening and what the market will charge for it. It is a grueling job, and those that succeed at it are the ones who truly keep the station on the air.

The station today is in many respects the same as it has always been. There has been no major shift in employees, and the voices and names that have been long associated with it still remain. Its alumnae include numerous Knoxvillians and East Tennesseans who went on to develop first-rate careers in broadcasting and other related fields. There was more than one colorful character to pass through the doors of WIVK. Radio is notorious for being a mix of everything in its employee base. The administration staff who came and went, the entertainers whose careers were started, the news broadcasters who went on to work in the big markets or serve as community leaders, announcers who rose to prominence here and elsewhere, and the advertising representatives that drove it all and advanced through the ranks all left their mark on the station. In spite of the good times, the open atmosphere necessary at any enterprise that feeds creativity, and the endless practical jokes that only good friends can pull on each other, Jim Dick ran a pretty tight ship and kept his hand on the wheel up until he retired from the business.

WIVK—the station that Jim Dick built—was sold to the Citadel Broadcasting Corporation in January 2001. The corporation is the third largest radio group in the United States, with a strong national footprint reaching more than 50 markets. It owns stations in eight of the top 10 markets in the United States. The company currently has 165 FM stations and 58 AM stations. WIVK is its leading one.

The station is still one of the resonating community voices. It leads in public affairs broadcasting and still boasts a vigorous news department that covers breaking events, features, and politics. With the advent of satellite radio like Sirius and XM, which offer music with no commercials, it seems like radio as Americans know it is on its way out. That may be the case in some cities, but in Knoxville, the radio market, though not what it used to be, is still strong and competitive.

The WIVK philosophies of Jim Dick are still working in that the radio is responsible to the community and its listeners, and provides what they want and need to hear in music, news, weather, sports, and traffic reporting. Announcers are still active in the community, and the station supports local and national charities. Even school groups on occasion come to visit. This last chapter is a look at WIVK then and now.

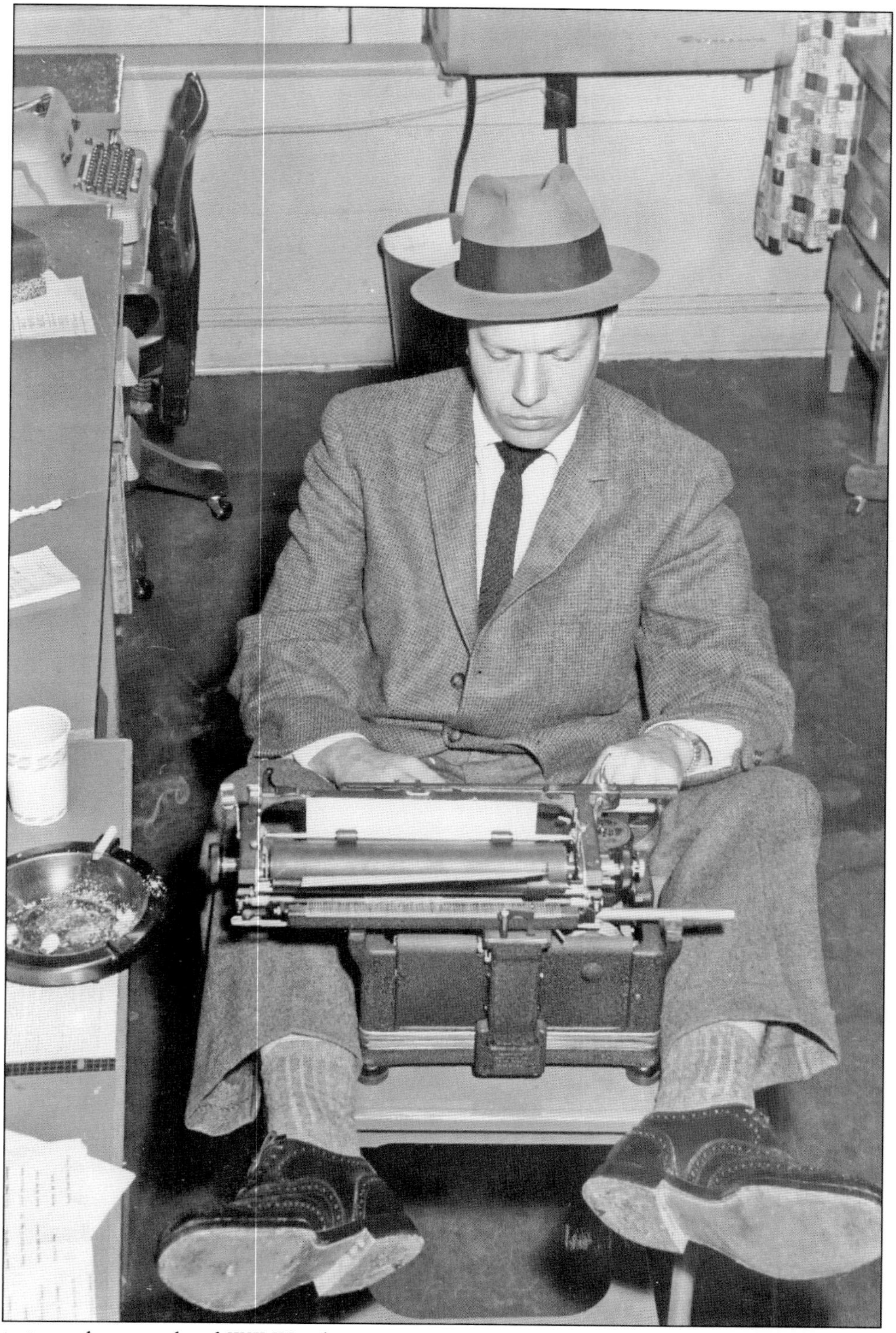

This is a photograph of WIVK advertising representative Ben Crenshaw, who used the air name Dan Todd, looking like the stereotypical radio guy with cigarette, coffee, and a typewriter between his knees. He was a notorious practical joker who gave Claude Tomlinson a run for his money.

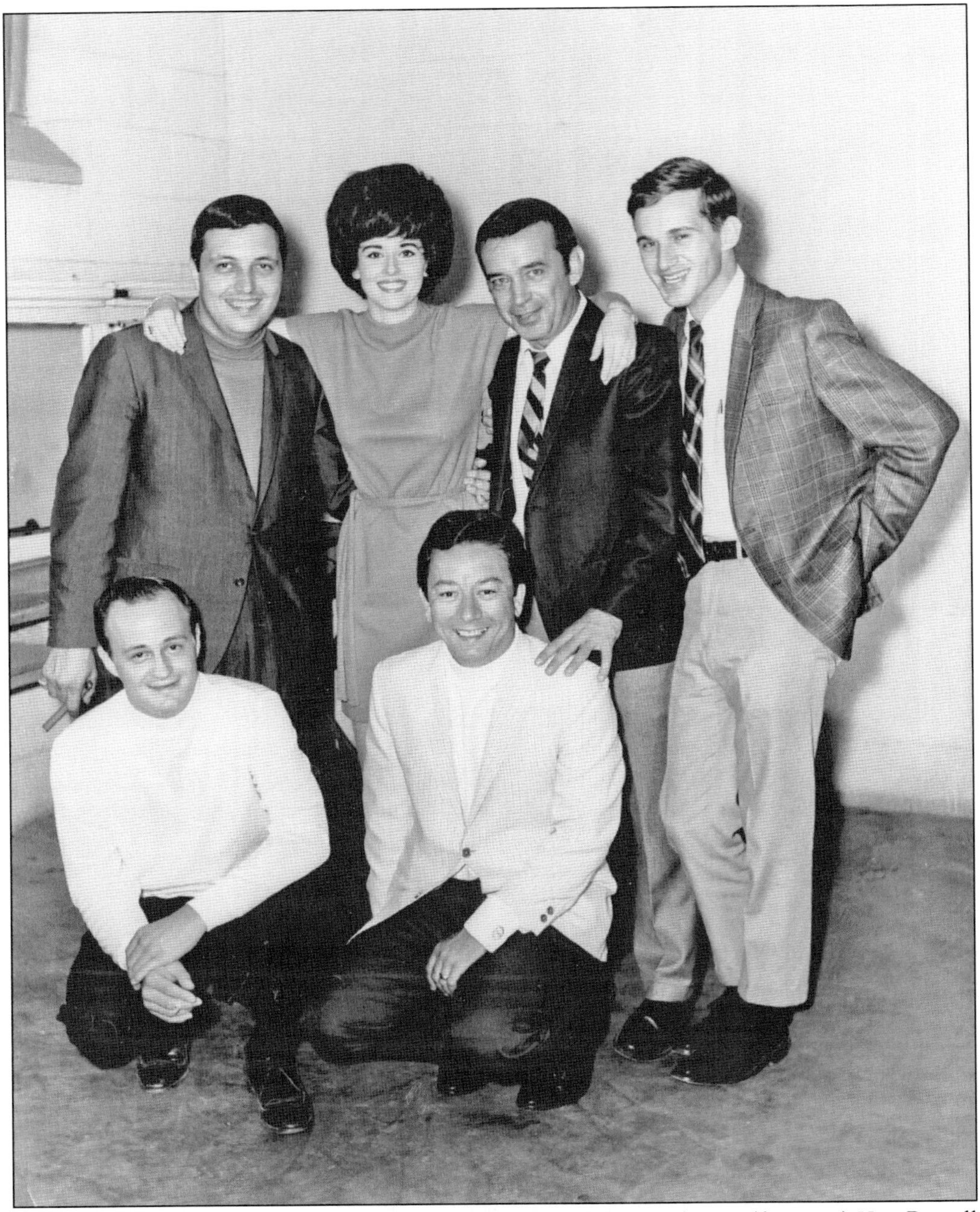

This is a photograph of the WIVK crew in 1968. From left to right are (first row) Ken Russell and Mike Hanes; (second row) Bobby Denton, Linda Gail Lewis (Jerry Lee Lewis's sister) Claude Tomlinson, and Ed Brantley.

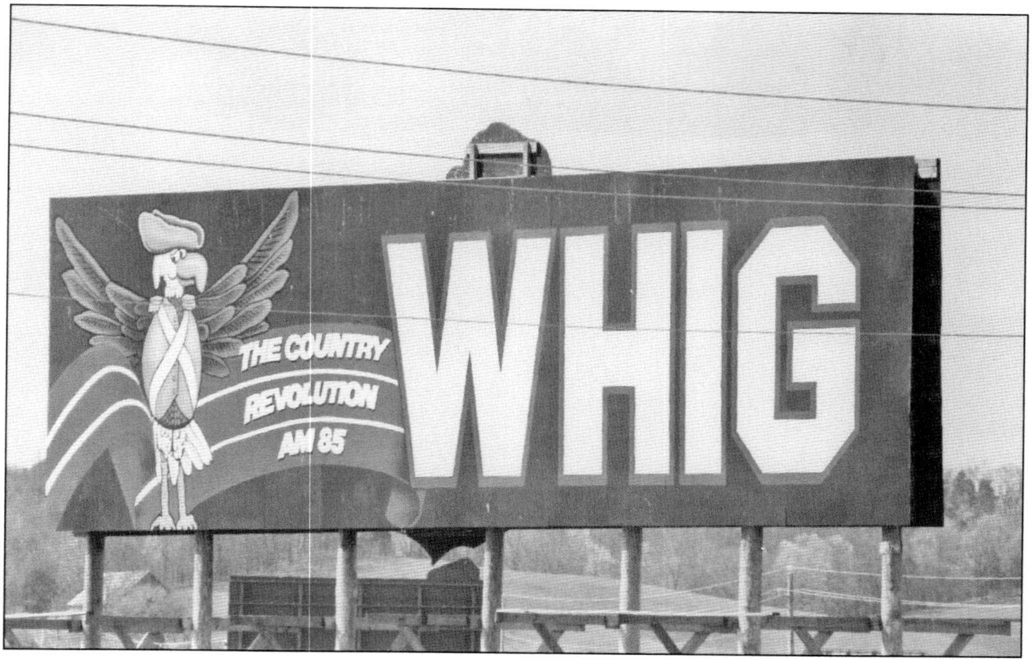

According to employees, there were only two mistakes Jim Dick made with WIVK. One was renaming the AM WHIG, which lasted only briefly before returning to WIVK AM. The other is seen below. This building was supposed to be WIVK's second home on Emory Road in Powell. He built it, but the station never moved there, instead staying at its home on Bearden Hill off Kingston Pike. The staff did not want to move to the new building, and the area reportedly had too many radio frequency waves for comfort. The building still stands.

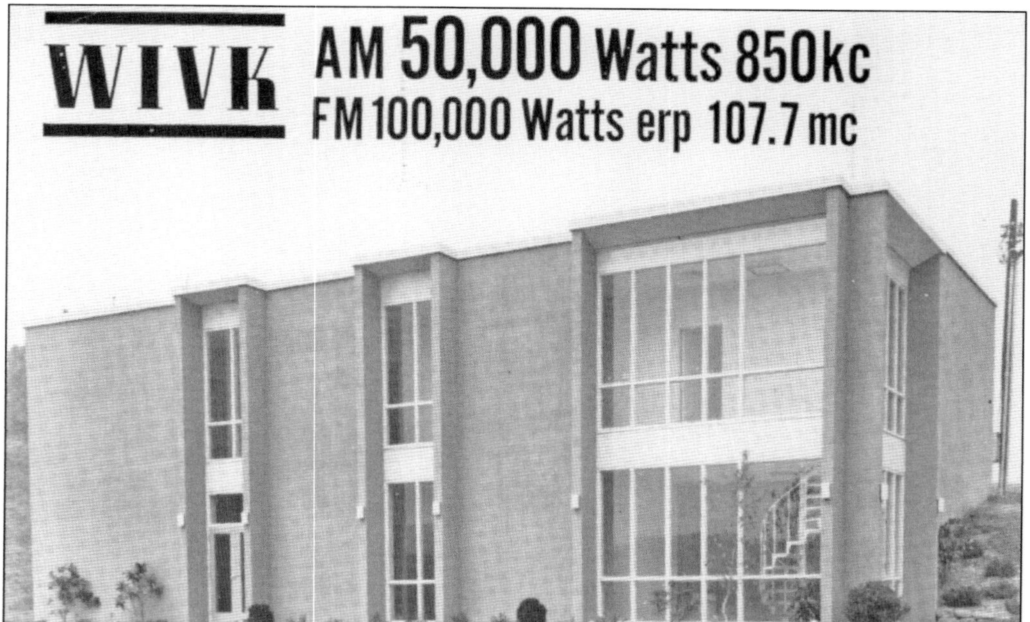

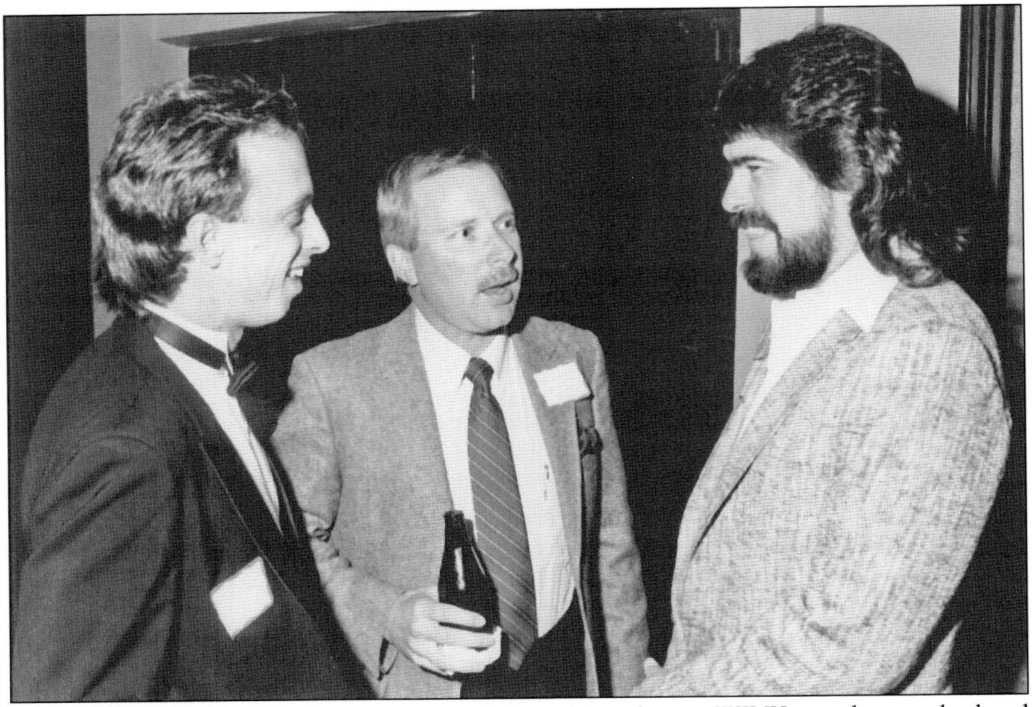

One of the most beloved bands to credit their rise to stardom to WIVK was the popular band Alabama. This image shows, from left to right, Mickey Dearstone, Norman Majors, and Randy Owen. The image below was taken at a listener appreciation concert in Knoxville. The group was playing a local nightclub in Knoxville. Alabama made frequent appearances at the station after the band's rise to stardom. (Photographs courtesy WIVK.)

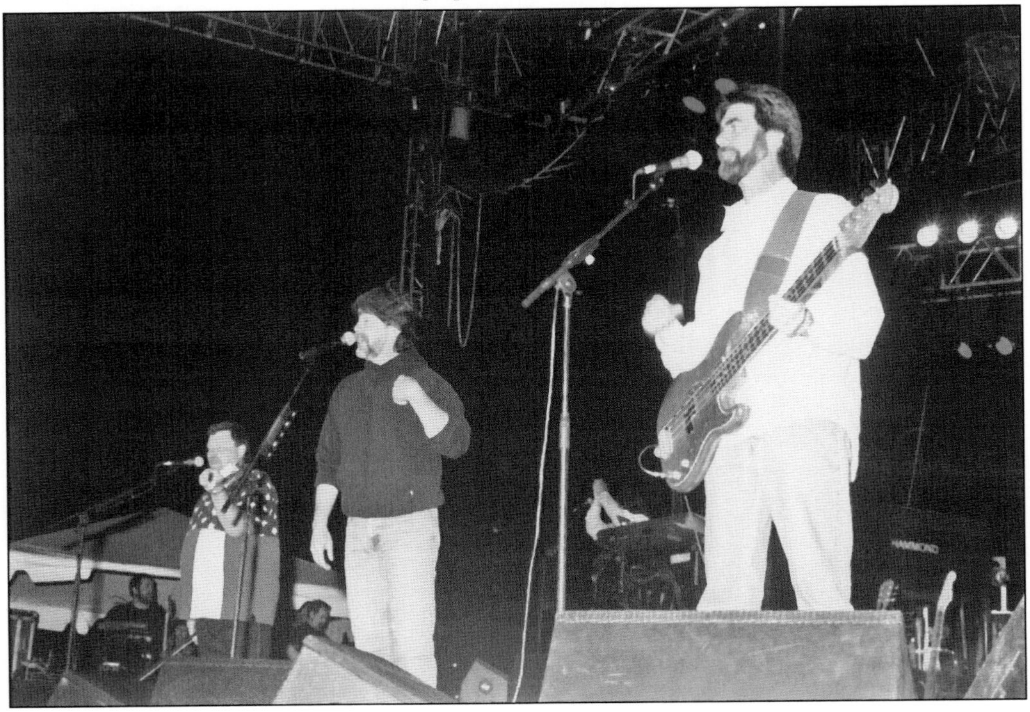

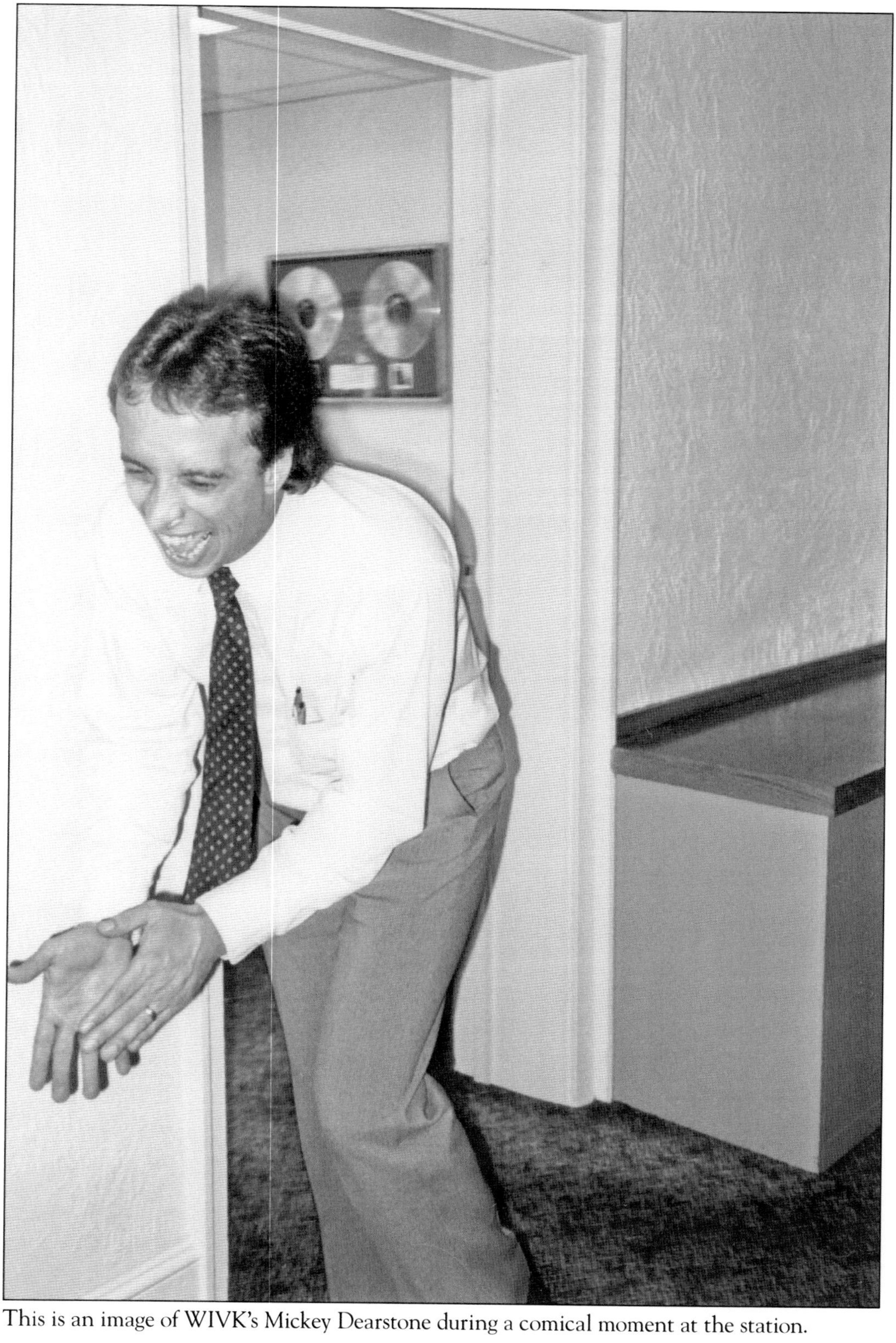

This is an image of WIVK's Mickey Dearstone during a comical moment at the station.

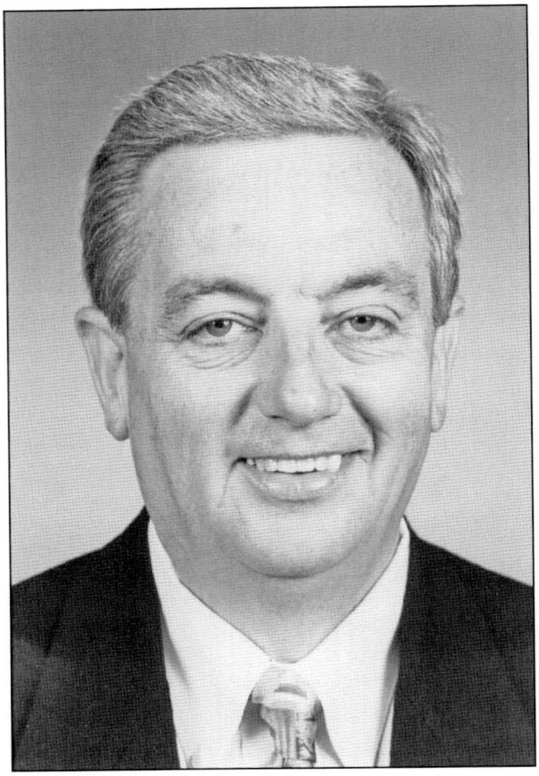

Les Acree was a WIVK program director in the 1990s. He hired Darren Wilhite and Tim Wall and the current morning team of Andy and Allison before leaving the station.

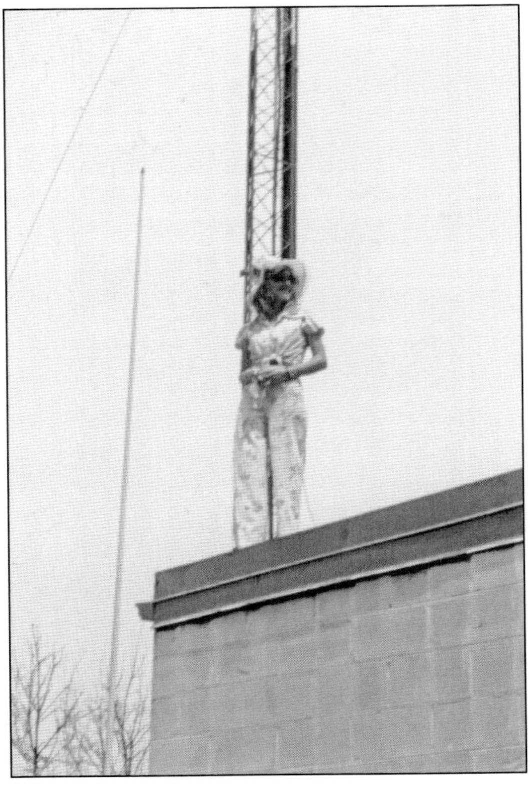

This is an image of Stella Parton—younger sister of Dolly—standing on top of the WIVK studio at 6711 Kingston Pike. The entire Parton family was part of WIVK's office family, and it is still common to see them today.

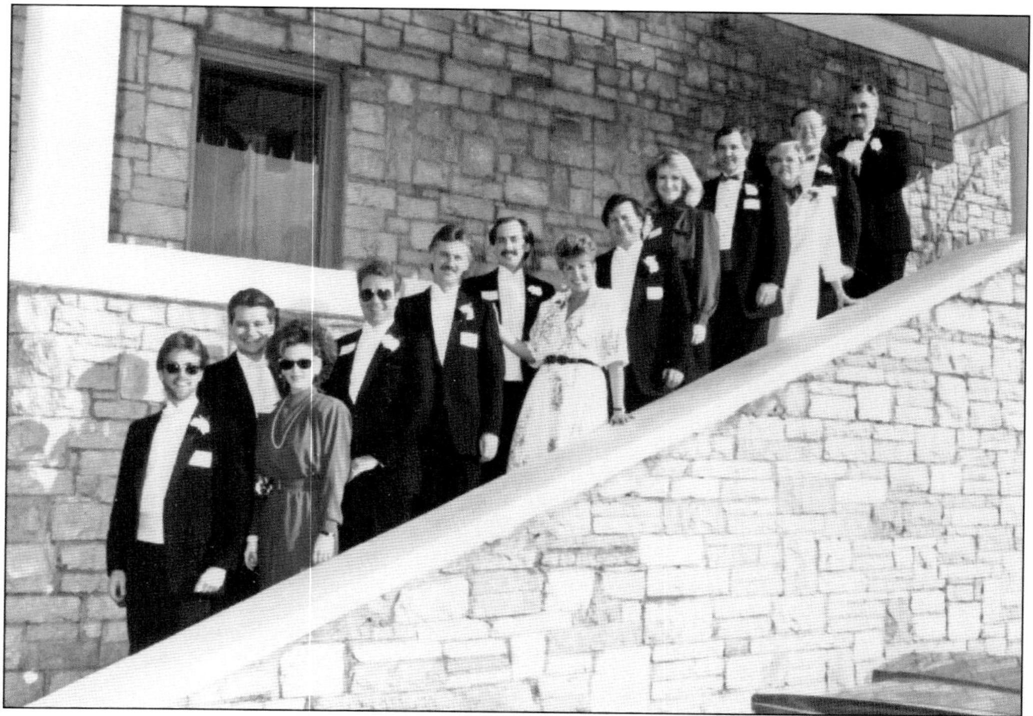

This is a group photograph of the sales and management staff at the WIVK 30th anniversary taken before a black tie function in Knoxville. The station held many such events through its history. (Photograph courtesy WIVK.)

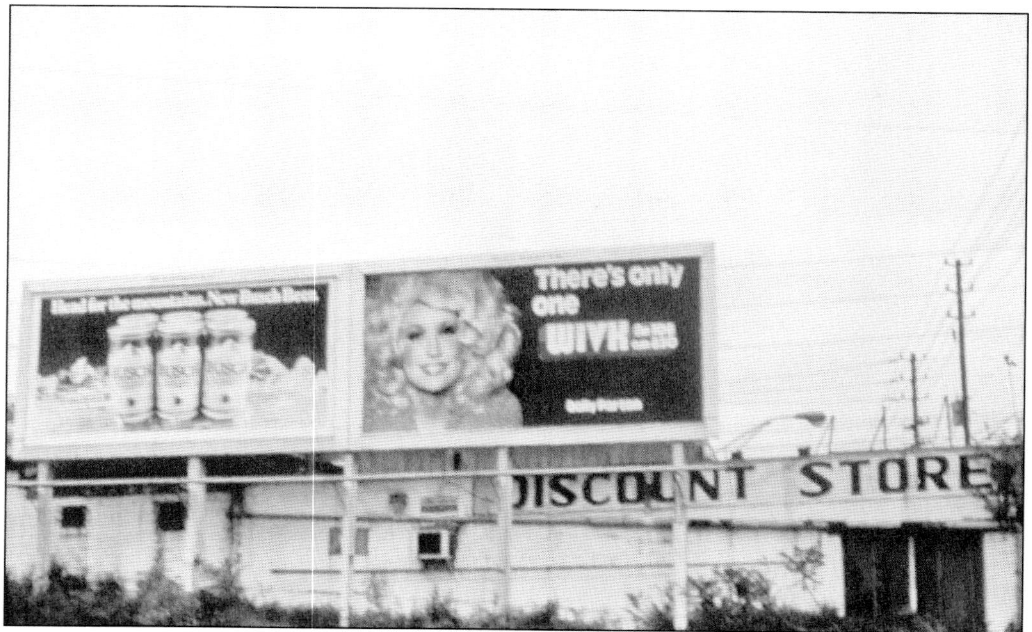

This billboard from the 1980s promotes the slogan "There's Only One," featuring local country music celebrity Dolly Parton.

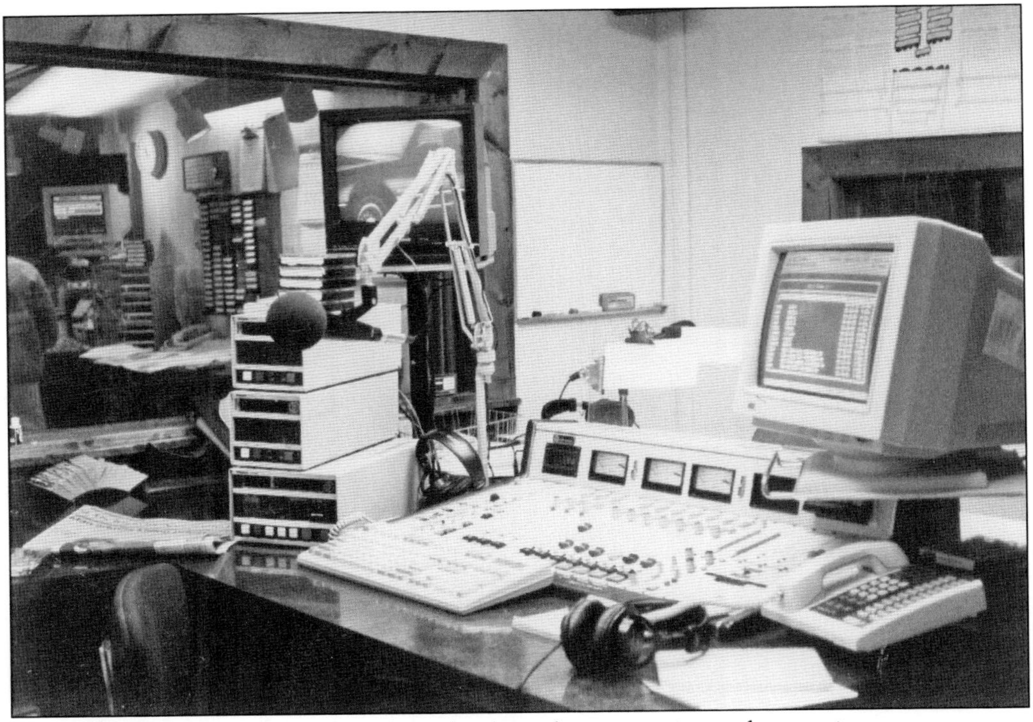

The WIVK newsroom studio went from the days of a wire service and typewriters to computers. This view still shows the old cart machines used for interviews. Today it is all done digitally. (Photograph courtesy Jean Ash.)

This photograph was taken in the 1980s before a black-tie function. From left to right are Gail Robeson, Tom Caulkins, Tony Igar, Linda Walker, Mike Hammond, Jenny Wylie, and Dave Young, and seated in the middle is Bobby Denton. (Photograph courtesy WIVK.)

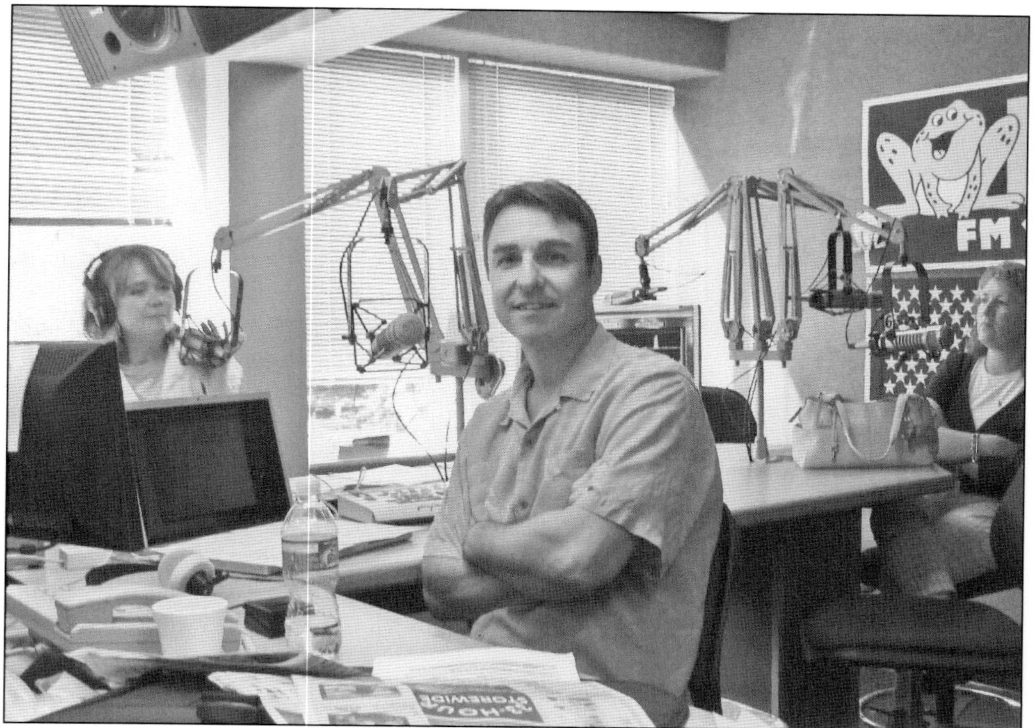

The Andy and Allison morning team was hired when Wilhite and Wall left the station. Pictured here are Andy and Alison doing their morning show. Below is their producer, Jimmy Holt. (Photographs courtesy WIVK.)

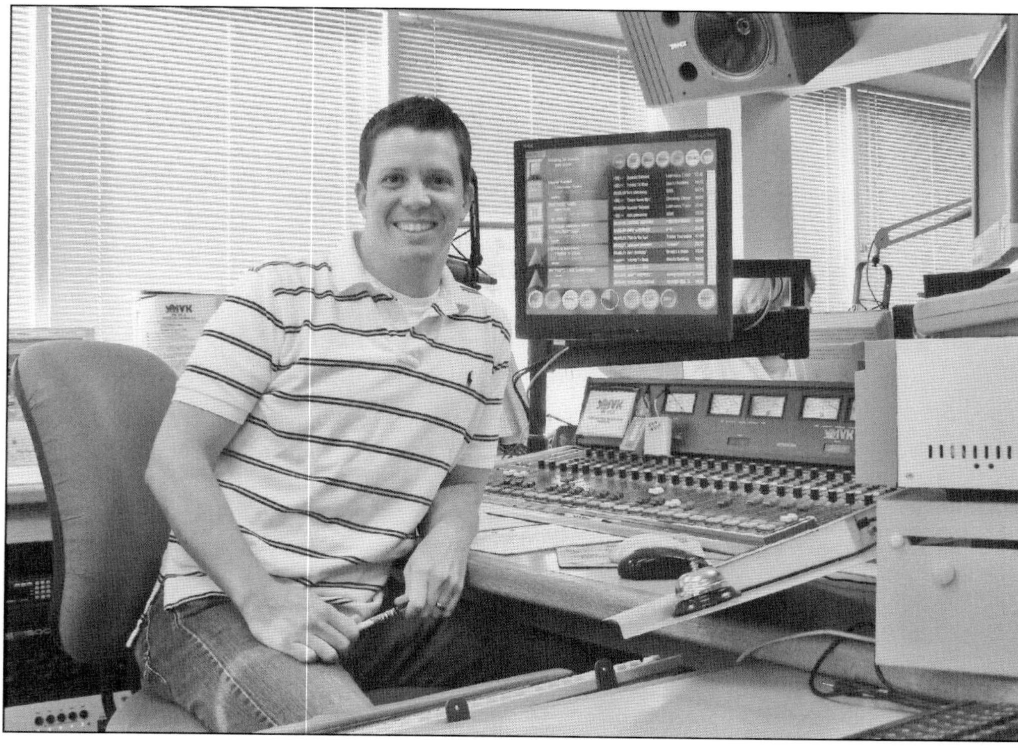

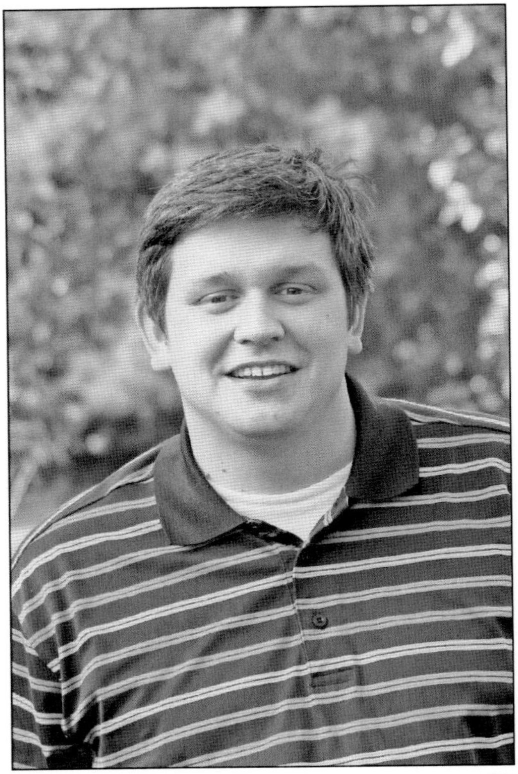

The WIVK morning team's sports anchor, Chip Kain, works with Andy and Allison and handles sports assignments for the station. (Photograph courtesy Chip Kain.)

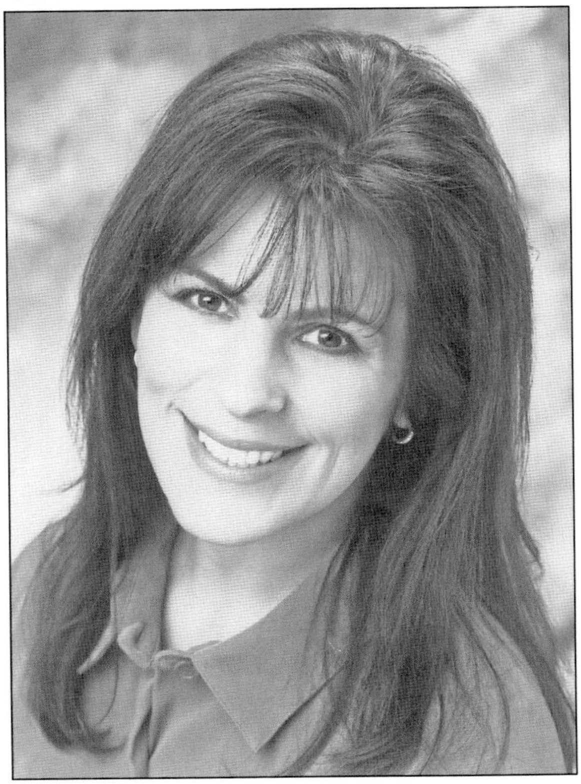

Colleen Addair is the popular midday voice for WIVK listeners. (Photograph courtesy WIVK.)

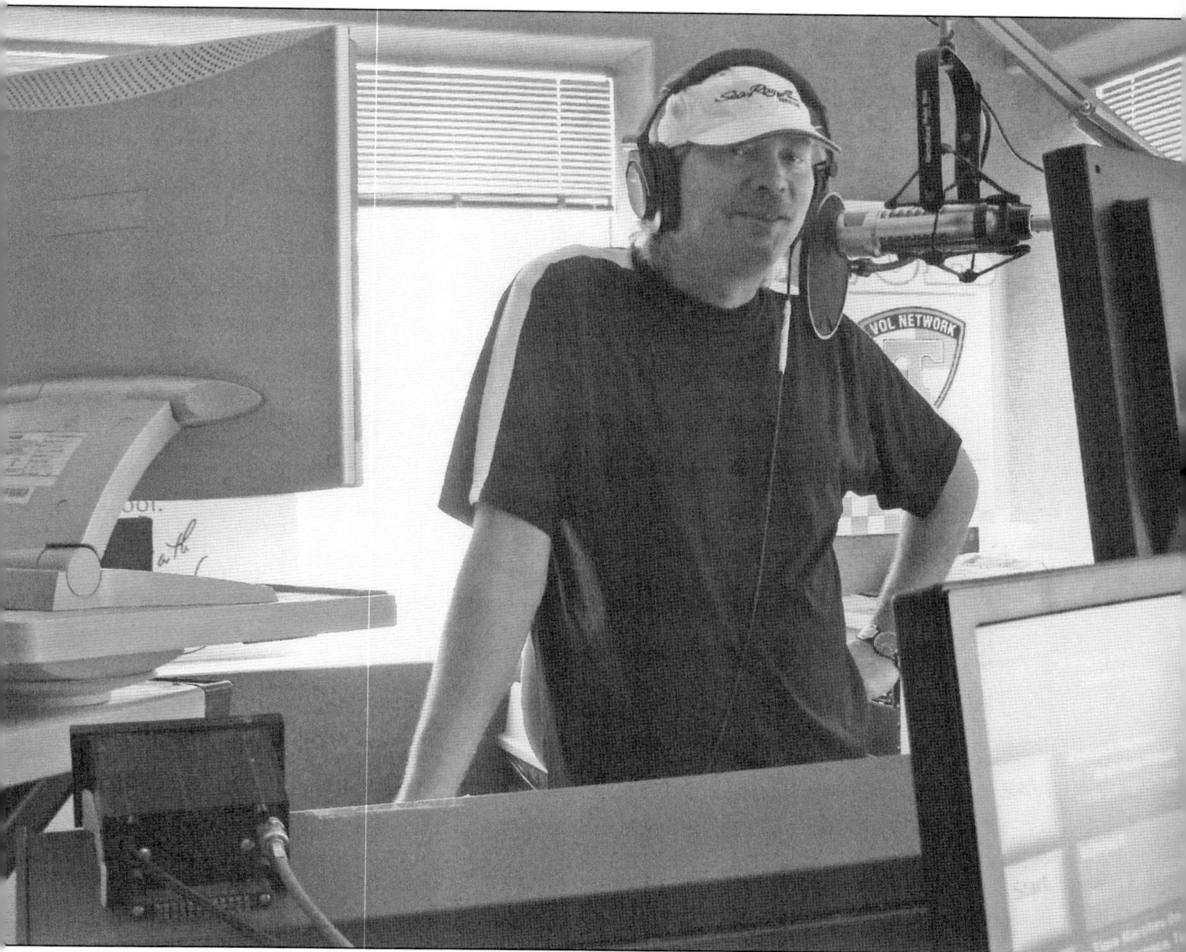

Afternoon drive announcer Ted "Gunner" Ousley is known for his "Allen Jackson" looks. Gunner's afternoon drive program is the most listened to in East Tennessee. (Photograph courtesy WIVK.)

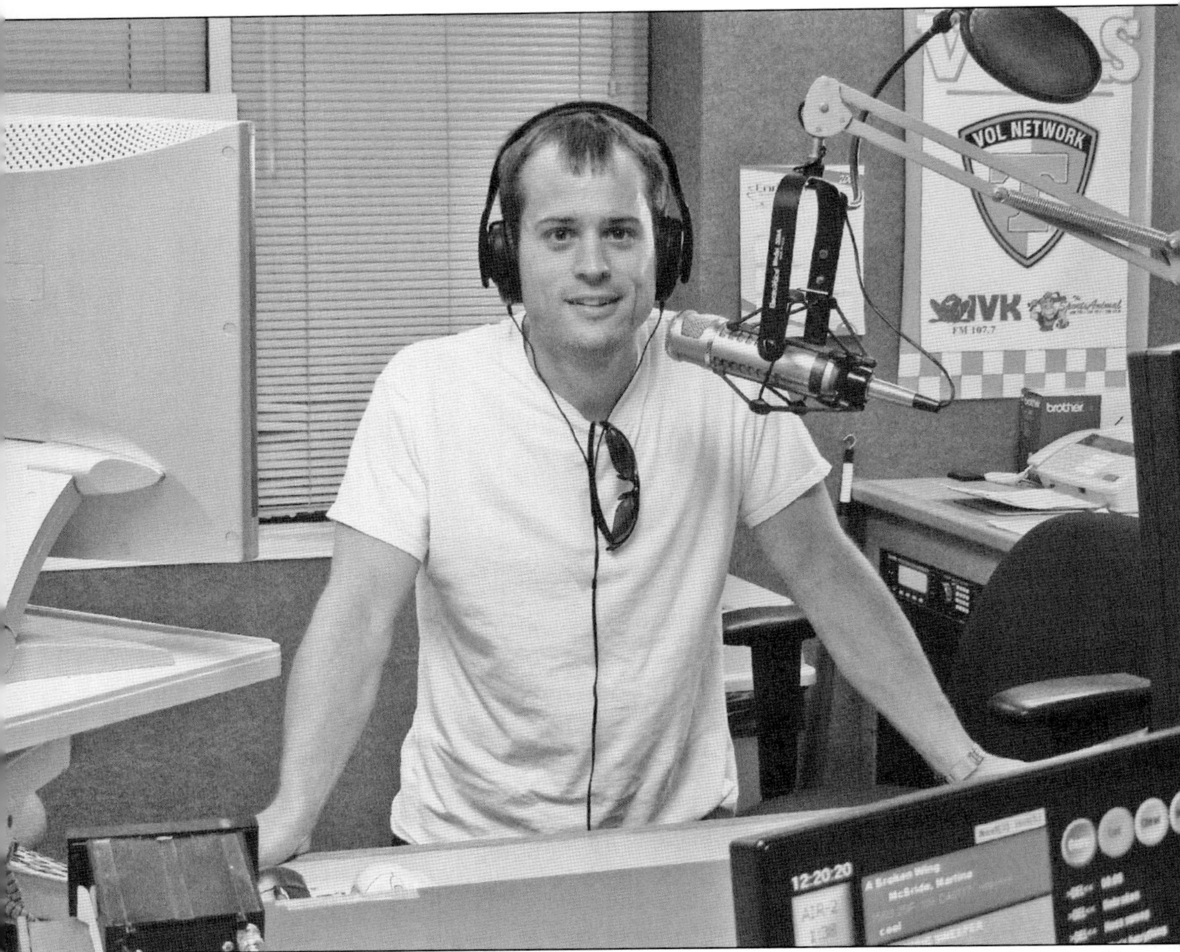

The nighttime announcer, Jack Ryan, grew up at the station. He has made a name for himself in the broadcast business in East Tennessee and is a popular nighttime talent. (Photograph courtesy WIVK.)

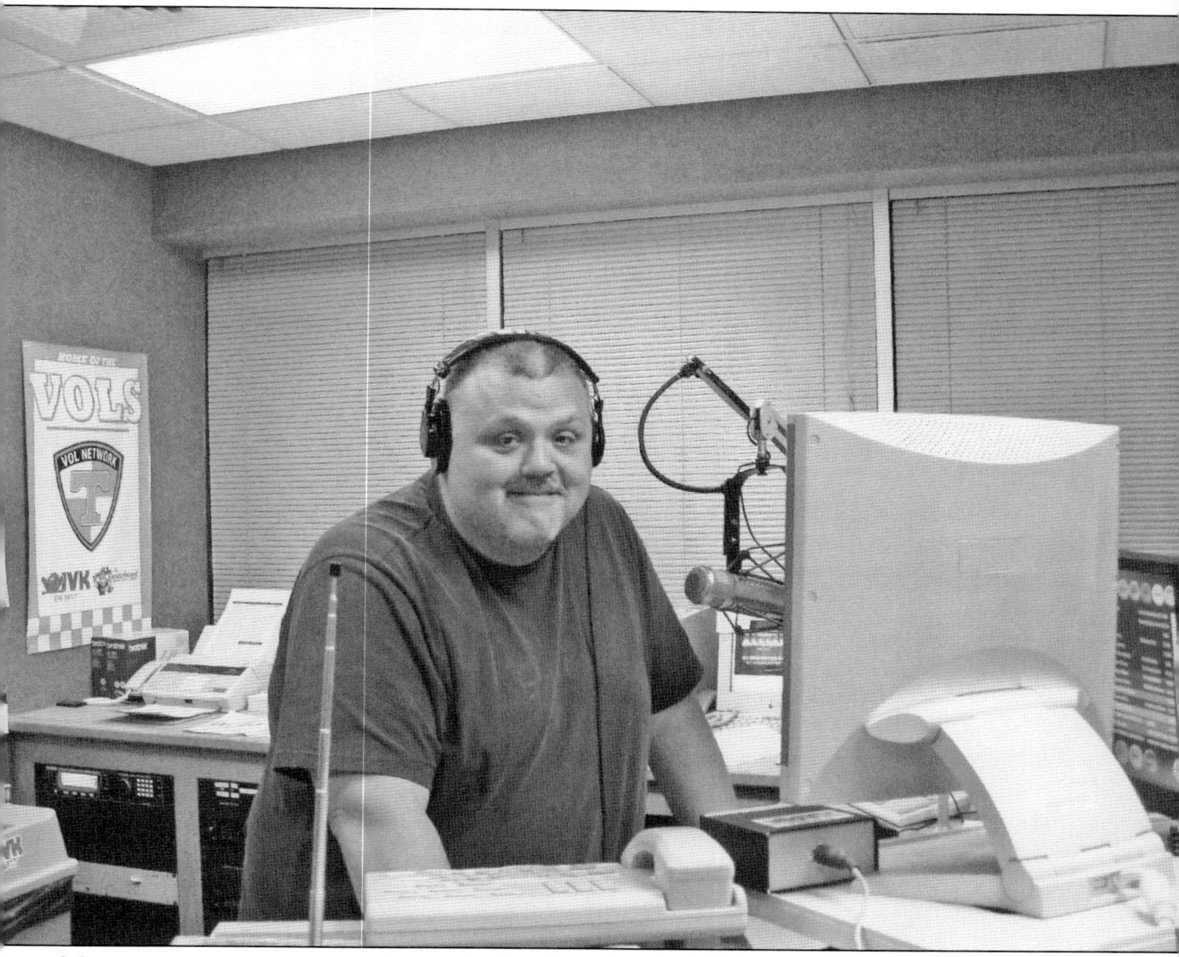

Morristown native Ron Cantwell is seen here. He is known as "Hoss" and does the graveyard shift from midnight to 6:00 a.m. (Photograph courtesy WIVK.)

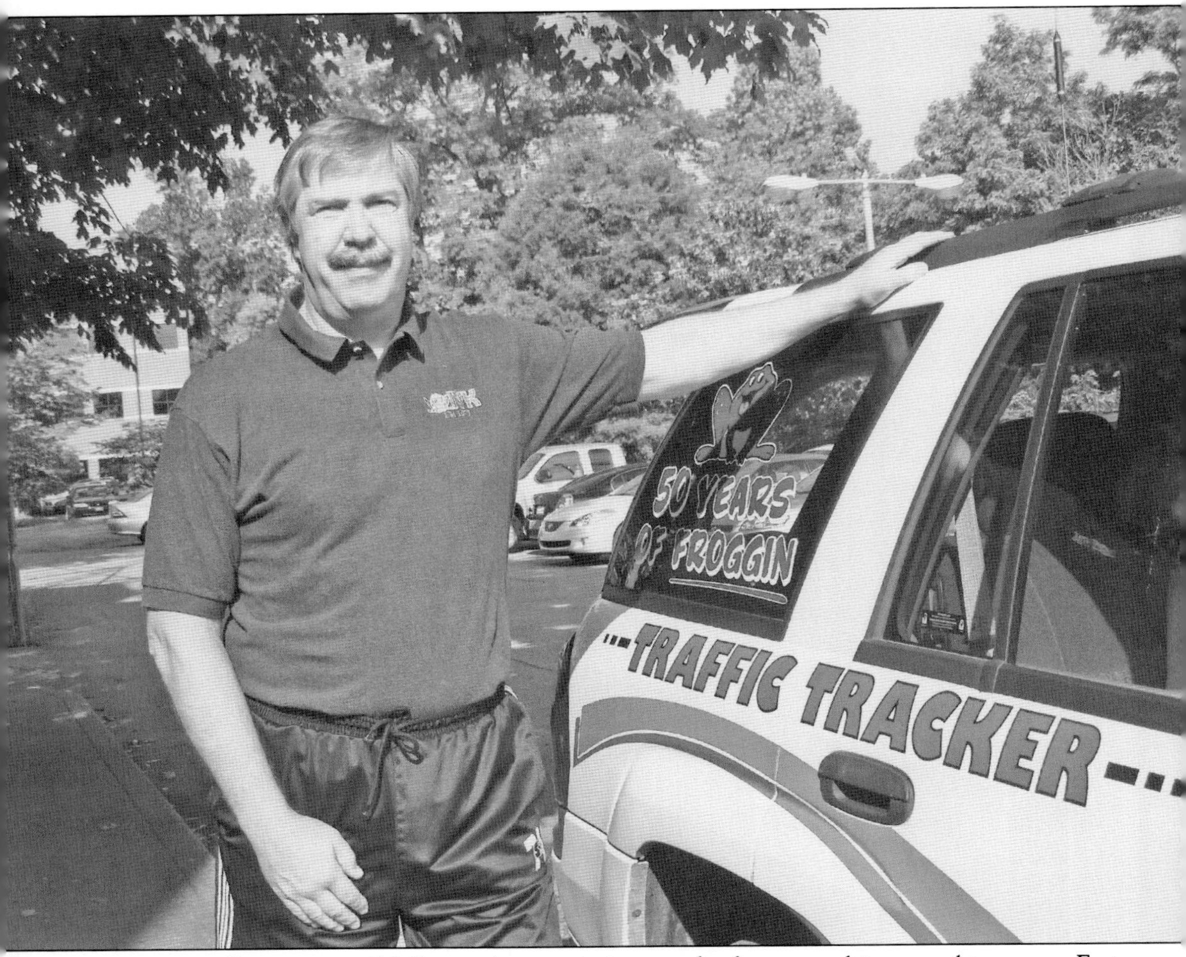

WIVK traffic reporter Ed Rupp does mornings and afternoon drives, making sure East Tennesseans get from home to work and back. (Photograph courtesy WIVK.)

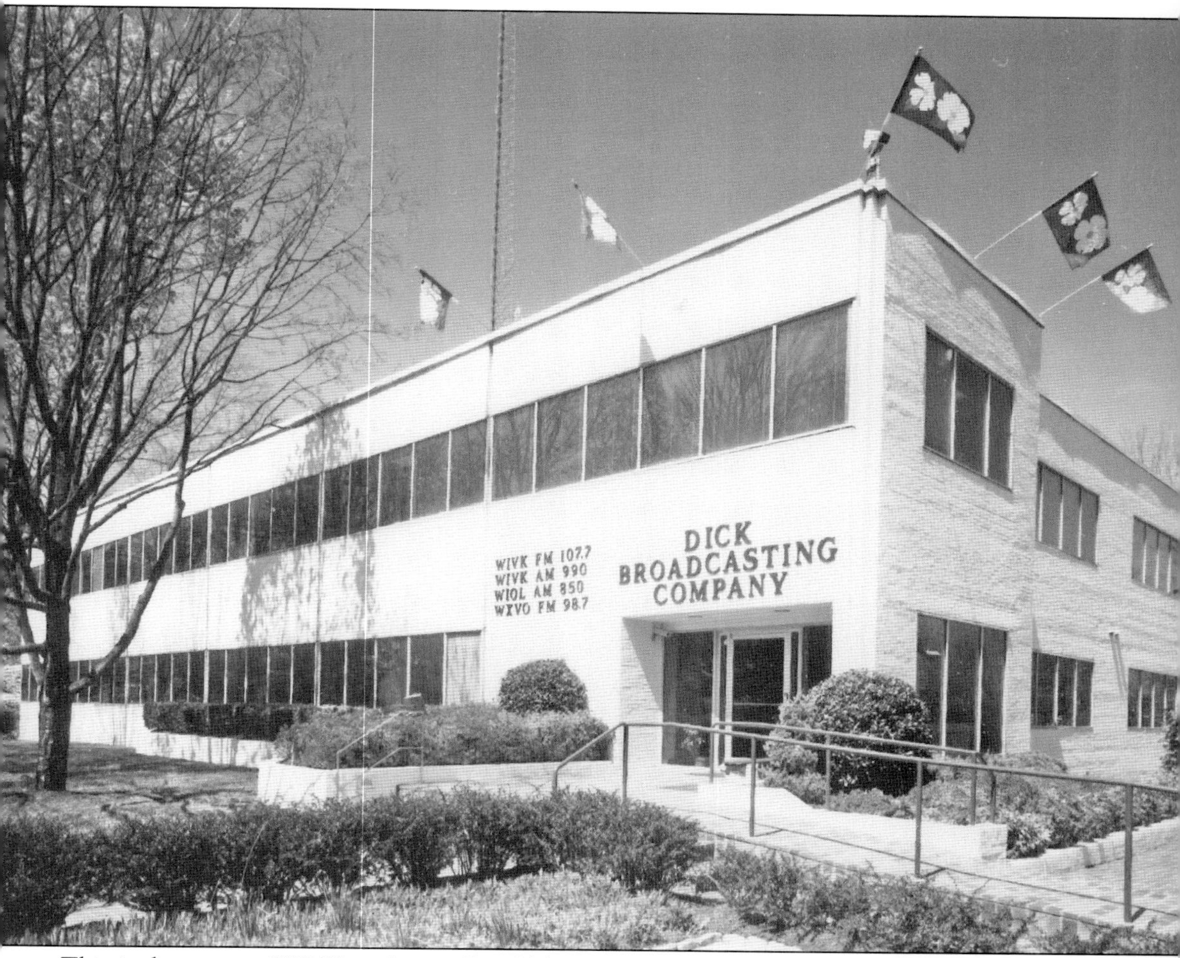

This is the current WIVK studio at 4711 Old Kingston Pike, which was bought by the Citadel Broadcasting Corporation in 2001. (Photograph courtesy WIVK.)

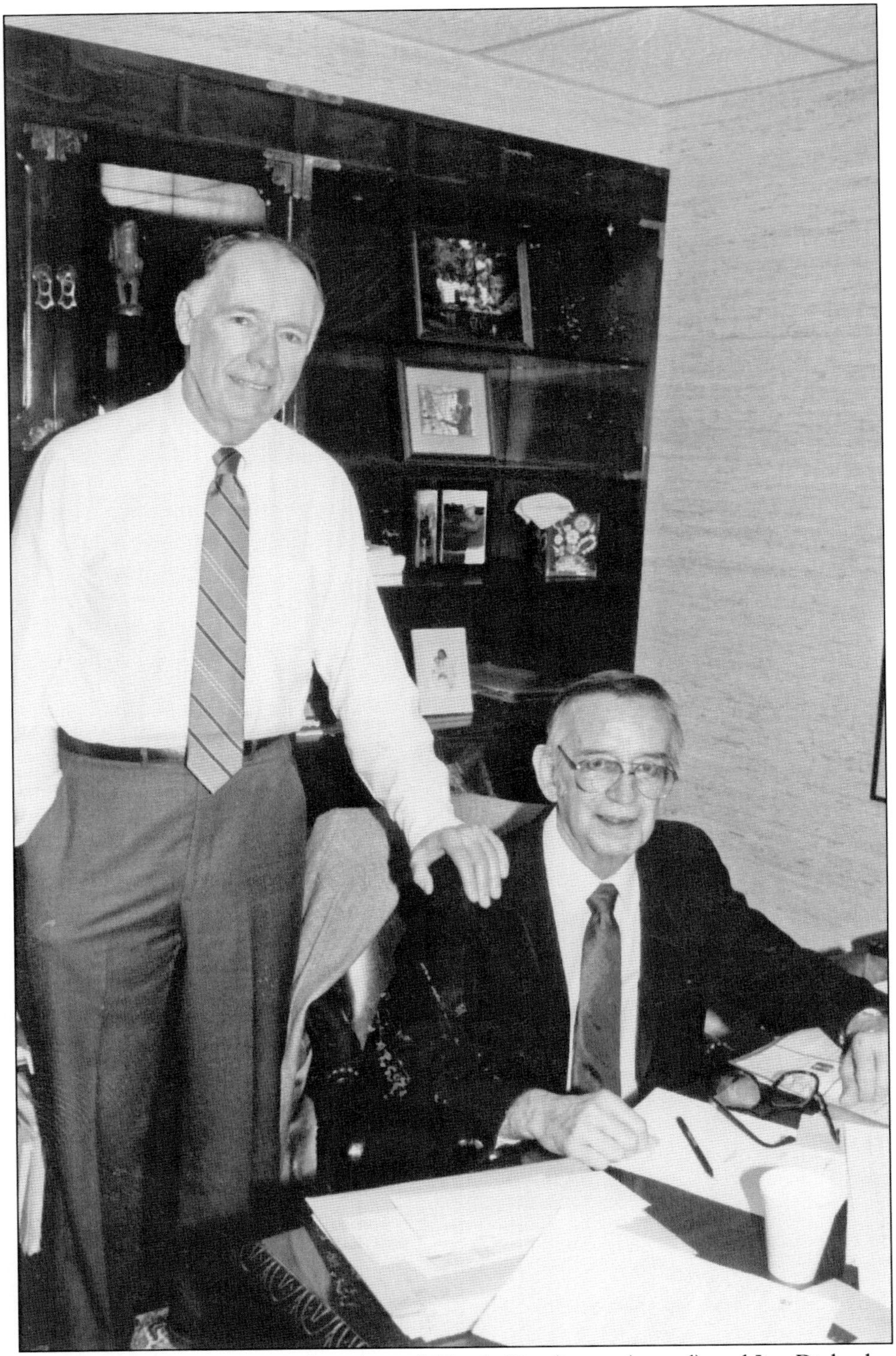

This is one of the last photographs taken of Claude Tomlinson (seated) and Jim Dick, the two men that started it all. It is only appropriate that this should be the last photograph in *Knoxville's WIVK*. The two men took a small little station on Knoxville's Gay Street and turned it into an American broadcast legend still on the air and going strong. (Photograph courtesy WIVK.)

BIBLIOGRAPHY

Deadrick, Lucille. *Heart of the Valley: A History of Knoxville, Tennessee.* Knoxville, TN: East Tennessee Historical Society, 1976.